MENINGITIS

SYMPTOMS, MANAGEMENT AND POTENTIAL COMPLICATIONS

BACTERIOLOGY RESEARCH DEVELOPMENTS

Additional books in this series can be found on Nova's website
under the Series tab.

Additional e-books in this series can be found on Nova's website
under the e-book tab.

BACTERIOLOGY RESEARCH DEVELOPMENTS

MENINGITIS

SYMPTOMS, MANAGEMENT AND POTENTIAL COMPLICATIONS

ANTHONY L. SHRADER
EDITOR

New York

Library of Congress Cataloging-in-Publication Data

ISBN: 978-1-63117-136-9

Library of Congress Control Number: 2013958461

Published by Nova Science Publishers, Inc. † New York

Contents

Preface   **vii**

Chapter 1   Drug-Induced Aseptic Meningitis:
A Challenging Diagnosis   **1**
Antonios Kerasnoudis

Chapter 2   Bacterial Meningitis: Procalcitonin Levels,
Initial Signs and Symptoms, and Complications
in a Reanimation Unit   **17**
*Carmen Mateo Revilla, Tomas Rodríguez Delgado,
Juan José Gómez Sainz
and Luciano Aguilera Celorrio*

Chapter 3   Tuberculous Meningitis: Update to an Old Disease   **31**
Fernando Alarcon and Gonzalo Dueñas

Chapter 4   Role of Prophylactic Antibiotics
in Posttraumatic Meningitis   **65**
Carlos Jimenez, Luisa Galvis and Manuela Jimenez

Chapter 5   Meningitis Associated with Autoimmune Diseases   **81**
Masakazu Nakamura and Masuaki Niino

Chapter 6   Sepsis and Bacteremia, Meningitis, Influenza
Infection and Infectious Diarrhea   **109**
Elif Doyuk-Kurtul and Hakan Erdem

Index   **123**

Preface

In this book the authors present current research in the study of meningitis. Topics discussed in this compilation include the drug induced aseptic meningitis; bacterial meningitis; the role of prophylactic antibiotics in posttraumatic meningitis; meningitis associated with autoimmune diseases; sepsis and bacteremia, meningitis, influenza infection, and infectious diarrhea.

Chapter 1 – Aseptic meningitis refers to a nonbacterial inflammation of the leptomeninges. Drug-induced aseptic meningitis is a rare, but important and often-challenging diagnosis for the primary care physician. Its incidence is unknown, while recent series have allowed for an approximation to the extent of this clinical problem. It has been well described in the literature, most frequently in association with medications such as nonsteroidal antiinflammatory drugs (NSAIDs), antibiotics, intrathecal medications, intravenous immunoglobulins.

One proposed pathomechanism is that the meninges are directly irritated by the in-trathecal administration of drugs, while an immunological hypersensitivity of the meninges (type 3 or type 4) to the offending drug may also play an important role. Signs and symptoms generally develop within 24 hours of drug ingestion including headache, fever, neck stiffness, mental status changes, malaise, nausea, vomiting, chills, generalized arthralgias, and myalgias. As the clinical examination may reveal classic signs, including nuchal rigidity, Kernig's and Brudzinski's signs, thus, differentiating this condition from infectious meningitis is crucial.

The diagnosis of this nosology is made by establishing a causal relationship between the use of the drug and the onset of signs and symptoms, supported by negative tests for infectious causes of symptoms and rapidity of resolution after the drug is discontinued.

This review article provides a concise summary of drug-induced aseptic meningitis, out-lining the challenges a primary care physician may face in making the clinical diagnosis.

Chapter 2 – The authors conducted a retrospective review of patients hospitalized in their reanimation unit during a period of time of 53 months (from January 2009 to May 2013), with the diagnosis of bacterial meningitis. The main goal of this chapter was to analyze the symptoms, APACHE II, SOFA and procalcitonin levels of these patients on admission and their evolution during their stay in the unit.

The authors evaluated 25 patients. Two of them were diagnosed nosocomial meningitis, and the rest community-acquired meningitis. 5 patients (20%) had smaller procalcitonin levels than 0.5 ng / dl on admission and did not increase during their stay in the authors' unit. The most common signs and symptoms were (from most to least): fever (84 %), altered level of consciousness (60 %), headache (48 %), neck stiffness (44 %), vomiting (40 %), ear pain (32 %), back pain (12%), neck pain (8 %), petechiae (8 %), photophobia (4 %), and seizure (4%). The most frequently isolated bacterium was pneumococcus in 14 patients (56 %), followed by Neisseria meningitidis in 3 patients (13.6 %) and Streptococcus pyogenes in 2 patients (9.1 %). Staphylococuss aureus, Escherichia coli, Haemophilus, Listeria and Bacteroides fragilis were each found in one patient. Most frequent complications were: requirement of mechanical ventilation (44%), coma (40%) and septic shock in (32%). Followed by rhabdomyolysis (20%), seizures (12%), pneumonia (8%), neurological deficit in (8%), urinary tract infection (4 %) and brain death (4%). Mortality rate in patients was 12 %. The patients who died had an APACHE II mean of 23 and SOFA of 7.67 on admission, compared with those discharged from the RU who had an APACHE II mean of 14.7 and SOFA 3.25.

Chapter 3 – Tuberculous meningitis (TBM) is the most severe form of tuberculosis. It is an important cause of death and disease among both children and adults throughout the world, especially in developing countries. The prognosis is worse in patients with HIV, in patients with severe neurological impairment, in chronic cases, and in patients with a resistance to drugs. The disease's pathogenesis has not as yet been sufficiently understood. Despite progress achieved over the past decades, a diagnosis of TBM is difficult to reach in many cases, because clinical features are non-specific and because there is not as yet any diagnostic test sufficiently sensitive, targeted and timely. The percentage of patients where a definitive diagnosis of TBM can be established is low. TBM diagnosis and treatment are a challenge for neurologists. There is no characteristic medical profile, which makes it difficult to reach a diagnosis. Systemic symptoms of infection by Mycobacterium tuberculosis (MT), along with meningeal signs, may be suggestive of tuberculous meningitis. HIV infection does not change how tuberculous meningitis is presented.

In 80% of patients with TBM, symptoms appear at least one week before diagnosis. The time profile for the onset of TBM symptoms could be used to differentiate tuberculous meningitis from bacterial meningitis. The percentage of patients diagnosed as having tuberculous meningitis might increase if a molecular diagnostic test were applied. Hydrocephalus, pre-contrast hyperdensity-hyperintensity, basal leptomeningeal enhancement, obliteration of the basal cisterns, infarcts and tuberculomas are highly frequent findingson CT and MRI studies. The diagnostic approach is based on clinical criteria, laboratory test results, and image findings.

Treatment delay is strongest risk factor for death. TBM prognosis depends on the presence of HIV, resistance to drugs, abnormal CT or MRI, substantial rise of proteins in CSF and onset of antituberculosis treatment three days after admission to hospital. The need to reach a diagnosis and administer early antituberculosis treatment is very important to reduce death and improve the prognosis. Severity on presentation is prognostic for outcome. Recently, a consensus in terms of clinical diagnostic, laboratory and imaging criteria have been defined and recommended for tuberculous meningitis to be used in future clinical research.

Chapter 4 – Traumatic brain injury (TBI) is one of the main causes of admission to neurosurgical services. The occurrence of meningitis after TBI, both closed and open, is rare but its early detection is very critical, since this infection increases morbidity and mortality up to 65%. Pathogens can access the central nervous system by direct contact and invasion through the external boundaries, or rarely from the bloodstream in association with a rupture of the blood-brain barrier. The conditions of barrier disruption are clear in penetrating trauma, and so is the pathogen colonization of the sinuses themselves and contact with the subarachnoid space in closed TBI by means of dural tears. The time elapsed from the moment of the trauma and the diagnosis of meningitis ranges from the first few hours to several years; the period has been reported from 8.4 days to 3.4 years. The mean interval of the different reports is 5-13 days, with 78% of cases diagnosed in the first 15 days. The microbiological agents isolated in post-traumatic meningitis include a wide range of both gram negative and gram positive bacteria, being Streptococcus pneumoniae the agent most commonly reported. The role of antibiotic prophylaxis in patients with CSF leaks has been extensively studied, yet still remains controversial. Published studies suggest that antibiotic prophylaxis does not reduce the risk of meningitis in TBI, both open and closed, since the odds ratio is not statistically significant in most of the clinical trials and prospective studies. The available scientific evidence does not support the use of prophylactic antibiotics to prevent posttraumatic meningitis, even though there are a few reports advocating its utility. When using prophylactic antibiotics, several factors must be considered: in addition to its unproven utility, it should be taken into account that the patient will be exposed to side effects that are not always mild, costs will increase and, last but not least, multiresistant microorganisms will be selected.

Chapter 5 – Autoimmune diseases can be organ-specific or systemic, with the latter type affecting many different organs. As a neurological complication particularly of systemic autoimmune diseases such as systemic lupus erythematosus and Behçet disease, meningitis can occur at any stage of the disease and can be a main symptom as well as a prognostic factor. Mechanisms of meningitis associated with autoimmune diseases are still unclear, although disease-specific autoantibodies and cytokines are considered to play critical roles in the pathogenesis.

In meningitis associated with autoimmune disease, the main symptoms are fever, headache, nausea, and neck stiffness, which are similar to those presenting in infectious, malignant, or other forms of meningitis. The clinical course is also varied and can be acute, chronic, or recurrent, and thus is not that helpful in differential diagnosis. It is extremely difficult to prove that comorbid autoimmune disease is a direct cause of meningitis, but the exclusion of other etiologies is essential for accurate diagnosis. Such diagnosis is made from a full set of physical and laboratory data and imaging findings. Furthermore, in patients with autoimmune diseases treated with immunosuppressive or immunomodulating agents, it is often difficult to differentiate meningitis associated with autoimmune disease from infectious or drug-induced meningitis.

Treatment options for meningitis related to autoimmune diseases are targeted to suppress activated autoimmunity, and immunosuppressive agents such as pulsed cyclophosphamide and steroids are often recommended for acute relapse or severe forms of the disease. Recently, new agents such as monoclonal antibodies have been tested in cases involving the central nervous system; however, their efficacy remains unclear.

In this chapter, the authors discuss autoimmune diseases with which meningitis can occur, and review the clinical features, diagnosis, and treatments for meningitis associated with autoimmune diseases.

Chapter 6 – Mortality in adults is frequently associated with sepsis, which becomes more common with advancing age. The mortality rate is five out of 100,000 in patients younger than 65 and increases to up to 26 for those over the age of 85. In elderly patients treated in the emergency department (ED), sepsis and bacteremia were reported to be involved in 18% of cases. In another report, positive blood cultures were obtained for approximately 10 % of the elderly patients in EDs. Escherichia coli was the most commonly isolated pathogen (29%), and the urinary tract was the most common source of infection (44 %).

In: Meningitis
Editor: Anthony L. Shrader

ISBN: 978-1-63117-136-9
© 2014 Nova Science Publishers, Inc.

Chapter 1

Drug-Induced Aseptic Meningitis: A Challenging Diagnosis

Antonios Kerasnoudis[*]
Neurological Department, St. Josef Hospital, Ruhr University,
Bochum, Germany

Abstract

Aseptic meningitis refers to a nonbacterial inflammation of the leptomeninges. Drug-induced aseptic meningitis is a rare, but important and often-challenging diagnosis for the primary care physician. Its incidence is unknown, while recent series have allowed for an approximation to the extent of this clinical problem. It has been well described in the literature, most frequently in association with medications such as nonsteroidal antiinflammatory drugs (NSAIDs), antibiotics, intrathecal medications, intravenous immunoglobulins.

One proposed pathomechanism is that the meninges are directly irritated by the in-trathecal administration of drugs, while an immunological hypersensitivity of the meninges (type 3 or type 4) to the offending drug may also play an important role. Signs and symptoms

[*] Antonios Kerasnoudis, MD. Neurological Department, St. Josef Hospital, Ruhr University, Bochum, Germany. Email: antonis.kerasnoudis@gmail.com. Tel: 004917662923235.

generally develop within 24 hours of drug ingestion including headache, fever, neck stiffness, mental status changes, malaise, nausea, vomiting, chills, generalized arthralgias, and myalgias. As the clinical examination may reveal classic signs, including nuchal rigidity, Kernig's and Brudzinski's signs, thus, differentiating this condition from infectious meningitis is crucial.

The diagnosis of this nosology is made by establishing a causal relationship between the use of the drug and the onset of signs and symptoms, supported by negative tests for infectious causes of symptoms and rapidity of resolution after the drug is discontinued.

This review article provides a concise summary of drug-induced aseptic meningitis, out-lining the challenges a primary care physician may face in making the clinical diagnosis.

Introduction

Aseptic meningitis refers to a nonbacterial inflammation of the leptomeninges. The causes of meningitis are either infective (bacterial, viral, fungal or protozoal) or noninfective. The noninfective (aseptic) causes include drugs, primary tumours of the central nervous system (CNS), carcinomas, granulomatous angiitis, sarcoidosis, systemic lupus erythematosus (SLE), rheumatoid arthritis, Behçet's syndrome, Vogt-Koyanagi-Harada syndrome and Mollaret's meningitis [1]. Although viruses are the most common cause of an aseptic meningitis, a drug-induced aseptic leptomeningeal infection (DIAM) may be a rare but often challenging diagnosis for the primary care physician.

The incidence of DIAM remains nowadays unknown. The first report of this nosology in the literature was in 1963, in a patient who had taken 2 tablets of sulfamethizole for his pancreatitis [2]. Several other reports have followed, mainly describing cases of DIAM after use of ibuprofen, antibiotics such as trimethoprim/sulfamethoxazole, immunoglobulines and monoclonal antiboides [3-12].

DIAM can mimic an infectious process, as well as meningitides that are secondary to systemic disorders for which these drugs are used. It is a diagnosis of exclusion, as clinical signs and cerebrospinal fluid findings vary greatly. The level of evidence regarding drug induced meningitis is largely in the form of anecdotal case reports and therefore must be interpreted carefully. Patients with DIAM typically present with fever, headache, and nuchal rigidity, usually within 24 to 48 hours after drug ingestion, but symptoms may occur until 2 years post-therapy [1]. CSF findings in DIAM vary considerably; however, in most of the cases there is a pleocytosis with hundred to several thousand cells per microlitre, elevated protein levels, whereas glucose is normal. Polymorphonuclear predominance is most common, while lymphocytic and eosinophilic findings have been also reported. By definition, CSF culture results are negative.

Considering the fact, that several from these drugs are commonly prescribed by primary care physicians, the purpose of this manuscript is to discuss the classes of causative agents, their potential mechanisms of action, predisposing factors to DIAM and its management.

Pathogenesis

Although the pathogenesis of drug-induced aseptic meningitis is not fully understood, two major mechanisms have been proposed. The first proposed mechanism is, that the meninges are directly chemically irritated by the intrathecal administration of drugs or contrast agent, usually as a result of the differences in lipid solubility, ionic strength, particle size and pH. The second one, which refers to nonintrathecal applied medication, is based on the assumption that drug-induced aseptic meningitis is an acute hypersensitivity reaction (most often type 3 or 4).

In these reactions, antibodies combine to form complexes with the drug (or metabolite) in the serum, activating complement (type III), or T cells reactive to the drug are recruited to a site of inflammation (delayed type hypersensitivity or type IV) [1]. Strong evidence of this theory are the temporal relationship between drug ingestion and development of symptoms, the progressively shorter incubation periods in recurrent cases, the development of classic hypersensitivity features, and the rapid resolution of symptoms after the drug is discontinued [1, 14].

It is interesting, that many connective-tissue diseases, such as systemic lupus erythematosus or rheumatoid arthritis, have been associated with DIAM [13-16]. The link between these 2 nosologies does not seem coincidental, given that many healthy individuals have used on steroid antiinflammatory drugs (NSAIDs) without developing drug-induced aseptic meningitis. This link is also consistent with a type III hypersensitivity mechanism, which specifically involves the formation of antibody complexes with the offending drug, and SLE is an immune complex-mediated disease.

Causative Drugs

1. Nonsteroidal Anti-Inflammatory Drugs

Nonsteroidal Anti-Inflammatory Drugs (NSAIDs) are one of the most common agents reported to cause an aseptic meningitis [17]. Among the NSAIDs Ibuprofen seems to be by far the most often described causative agent in the literature, even in a single dosage of 200-400mg [17-20] (Table 1). Ibuprofen induced aseptic meningeal infection seems to occur more often in patients with known connective-tissue diseases [1, 17, 18], while reports on previously healthy subjects do already exist [21]. The role of immune complexes has been proposed from several authors as critical for the genesis of ibuprofen induced DIAM [22-24], mainly based on the study of Chez et al., who found an 8.75-fold increase in CSF-immune complexes over the normal serum values in a healthy patient with ibuprofen-induced meningitis [25]. However these findings were not confirmed from other study groups [26]. Based on the above findings, some authors have concluded, that ibuprofen-induced meningitis is an antigen-specific humoral immune process confined to the central nervous system, where the drug, and not a metabolite, would potentiate the activity of a preexisting autoantibody, resulting in complement fixation and development of an acute meningitis [1]. In addition to the common signs and symptoms of aseptic meningitis, lethargy, seizures, confusion, periorbital oedema, diplopia, conjunctivitis, hypotension, parotitis, pancreatitis have been also reported in ibuprofen related DIAM [18, 19].

Other NSAIDs that are causaly associated with DIAM are naproxen, Sulindac, diclofenac, piroxicam and ketoprofen [27-35] (Table 1).

2. Antibiotics

The reporting of antimicrobial-associated DIAM is complicated by previous antimicrobial therapy, since patients may receive several different agents in the treatment of an infection. The mechanism of action of antibiotic-induced meningitis is also supposed to be due to a hypersensitivity reaction for the same reasons described for NSAIDs. Unlike NSAIDs, immune complexes have been detected in the serum of 3 patients with meningitis induced by trimethroprim-sulfamethoxazole–induced meningitis, but these were not found in their CSF [36, 37]. In a case of cephalosporin-related DIAM, immune complexes were present both in serum and CSF, as well as an increased IgG value in the CSF. This may suggest either that transfer of immune complexes to the CSF is not involved in the pathogenesis of the illness or that they are not accessible for assay [38]. Joffe et al. suggest that immune complexes consisting in part of trimethroprim-sulfamethoxazole may have a predilection for deposition in the choroid plexus, inducing a necrotizing small-vessel vasculitis and leading to an aseptic meningitis [36]. On the other side, Gordon et al. found high levels of immune complexes in the CSF, suggestive of an intrathecal immune response [39].

Table 1.

Medications	Common	Uncommon	Rare
NSAIDs	Ibuprofen	Naproxen	Ketoprofen
		Diclofenac	Piroxicam
		Sulindac	Salicylates
			Tolmetin
Antibiotics	Trimetophrim	Sulfonamides	Cephalosporins
	Trimetophrim/ sulfamethoxazole		Amoxicillin
			Penicillin
			Isoniazid
			Ciprofloxacin
			Metronidazole
			Pyrazinamide
			Valacyclovir
Immunomodula ting agents	IVIG	Azathioprine	Infliximab
	monoclonal antibody OKT3		Sulfasalazine
Intrathecal agents		Metrizamide	Methotrexate
		Cytarabine	Gentamicin
		Methylprednis olone acetate	Hydrocortisone
			Baclofen
			Gadolinium
			Diethylenediamine pentaacetic acid
			spinal anesthesia
Miscellaneous		Carbamazepine	pyrazinamide
			ranitidine
			phenazopyridin
			levamisole
			indinavir

Cotrimoxazole (trimethoprim-sulfamethoxazole) is the most common antibiotic associated with aseptic meningitis [40-44]. The association of cotrimoxazole with autoimmune disease is still evident, although not as strong as with NSAIDs [45, 46]. Cephalosporins have been implicated in a woman who had several episodes of aseptic meningitis associated with exposure to cefalexin, cefazolin and ceftazidime [47], while Penicillin has been reported as causing meningitis in a woman after parenteral administration [48]. Isoniazid, ciprofloxacin, metronidazole, Gentamicin have been also reported as a cause of DIAM [49-51] (Table 1).

3. Intravenous Immunoglobulin

Intravenous immunoglobulin (IVIG) is a well described cause of aseptic meningitis [53-56]. 3,6- 12% of the patients were reported to present with symptoms of DIAM after receiving IVIGs in the concept of various studies (Table 1) [57,58]. High dose of IVIG seems to play an important role in the risk of DIAM and other adverse effects, such as renal failure and heamolytic anemia [58]. The possibility that this type of DIAM is caused by sensitivity to the stabilizing agents of the commercial preparations, such as polyethylene glycol, maltose, sucrose, or glycine, seems unlikely since this syndrome developed in the same patients who received other prior immunoglobulin preparations [54, 59]. Considering the presence of eosinophils in the CSF of some patients, Sekul et al. have suggested a hypersensitivity reaction caused by the direct entry of the immunoglobulins into the cerebrospinal compartment [58]. The same authors also proposed an alternative mechanism involving IgG, as it was increased in all the patients in whom it was measured, although IgG indexes did not reveal intrathecal synthesis. Another possible mechanism could be the possibility of IgG can entering the CSF through a blood-brain (or blood-nerve) barrier breakdown. Since the infused IgG, derived from a pool of more than 5000 donors, is allogenic, it could interact with antigenic determinants on the endothelial cells of the meningeal vasculature, resulting in a cytokine-mediated inflammatory reaction [58, 59].

Therefore, a number of measures are needed to prevent or reduce the problems associated with high dose IVIG [60]. The initial infusion should be given at a slow rate (not faster than 6 g/h) and at a 3% dilution. If this infusion is uncomplicated the rate and concentration may be increased. Prehydration is important and patients are encouraged to maintain a good fluid intake throughout treatment. Paracetamol (acetaminophen) alone or with codeine is used as premedication and in some cases antihistamines (cetirizine) have been of benefit. A number of reports of aseptic meningitis describe eosinophils in the CSF, and cetirizinemay have effects on their migration in addition to H1 blockade [60, 61]. The symptoms of aseptic meningitis are not always recurrent and treatment can often be continued if appropriate action is taken. In our experience, corticosteroids have not been of great benefit in the management of high dose IVIG-induced aseptic meningitis, and symptoms may occur even when patients are receiving high dosages of corticosteroids prior to the introduction of high dose IVIG.

4. Intrathecal Medication

There are various reports on intrathecal drugs or radiographic contrast media, that may cause a direct meningeal irritation and the clinical signs of an aseptic meningitis.

The toxicity or likelihood of irritation is related to concentration of the drug or contrast agent, particle size, lipid solubility, ionic strength and duration of contact with the cerebrospimal fluid.

Metrizamide has been reported as the most common contrast media agent causing DIAM after intrathecal application (Table 1) [62, 63]. Similar reports exist though after application of indium-diethylenetriaminetetra-acetic acid during cysternography, or lophendylate [64, 65]. On the other hand, only one case report exists concerning a DIAM after application of magnetic resonance imaging contrast agent gadolinium diethylenediaminepenta-acetic acid [66].

Concerning intrathecal drug application, hydrocortisone and bothmethylprednisolone acetate are the most common drugs, that have been reported in the literature as a cause of aseptic meningitis following intrathecal administration [67-71]. In addition, the intrathecal application of chemotherapeutic agents, such as methotrexate, cytarabine or baclofen may also cause a DIAM [72-74]. On the other hand, spinal anaesthesia has also been associated with aseptic meningitis, and contaminants within the preparation injected, such as starch from surgical gloves or phenolic disinfectant [75-76].

5. Vaccines

Aseptic meningitis after vaccination is a rare complication. Vaccines associated in the literature with aseptic meningitis include monovalent mumps and rubella or polyvalent measles, mumps and rubella [77-80]. On the other hand, although aseptic meningitis following immunisation against hepatitis B vaccine has been reported, there are no existing reports of DIAM after diphtheria, tetanus and pertussis (DTP) vaccine [81].

6. Miscellaneous Drugs

There are various drugs associated in the literature with aseptic meningitis, among which carbamazepin, azathioprine, lamotrigine are the most common ones [82-85]. Monoclonal antibodies against CD3 (anti–T cell antibodies) causing aseptic meningitis have been the subject of many case reports [86, 87]. Other drugs associated rarely with DIAM include pyrazinamide, ranitidine, phenazopyridine, sulfasalazine, levamisole, indinavir and radiolabelled albumin [88-93].

Diagnosis – Therapy

Drug induced aseptic meningitis remains a diagnosis of exclusion. It is important to obtain a history of medical disorders such as systemic lupus erythematosus, the most frequent underlying condition associated with drug-induced aseptic meningitis. It is also important to make inquiries about recent vaccinations that may be implicated in the development of aseptic meningitis or exposure to other causative agents, recent lumbar puncture, or viral infections. A careful documentation of the drug history of the patient and their relationship to the onset of symptoms, previous reactions to the suspected agent or similar drugs, and recurrence of symptoms with rechallenge.

When DIAM is suspected, the drug should if possible be discontinued, and any infection risk should be promptly covered with appropriate antimicrobial such as a third generation cephalosporin, until a negative CSF infection screen is confirmed. In the case of high dose IVIG, fluid intake should be encouraged and intravenous saline may hasten the resolution of headache if this is already established. If further high dose IVIG treatments are required, premedication with a combination of antihistamines, paracetamol and hydration, and a slowIVIG infusion rate, may prevent or lessen symptoms.

Conclusion

DIAM has been reported as an uncommon adverse reaction with numerous drugs and chemicals and should be considered in the differential diagnosis of acute and recurrent meningitis. The major categories of causative agents are NSAIDs, antimicrobials, IVIG, intrathecal agents and vaccines, with a number of other agents reported less frequently. The association between SLE and ibuprofen as cause of DIAM is particularly strong and is important to recognise, as this medication may be purchased over the counter. DIAM is treatable by withdrawal of the drug and does not usually result in long term sequelae, even where repeated exposure to the drug has occurred.

References

[1] Jolles S, Carrock Sewell WA, Leighton C. Drug-Induced Aseptic Meningitis Diagnosis and Management. *Drug Safety* 2000 Mar; 22 (3): 215-226.

[2] Barrett PV, Thier SO. Meningitis and pancreatitis associated with sulfamethi-zole. *N Engl J Med.* 1963;268:36-37.

[3] Davis BJ, Thompson J, Peimann A, et al. Drug-induced aseptic meningitis caused by two medications. *Neurology.* 1994;44:984-985.

[4] Jolles S, Sewell WA, Leighton C. Drug-induced aseptic meningitis: diagnosis and management. *Drug Saf.* 2000;22:215-226.

[5] Muller MP, Richardson DC, Walmsley SL. Trimethoprim-sulfamethoxazole in-duced aseptic meningitis in a renal transplant patient. *Clin Nephrol.* 2001;55:80-84.

[6] Kato EShindo SEto Y ct al. Administration of immune globulin associated with aseptic meningitis. *JAMA* 1988;2593269- 3271.

[7] Casteels-Van Daele MWijndaele LHunninck KGillis P Intravenous immune globulin and acute aseptic meningitis. *N Engl J Med.* 1990;323614- 615.

[8] Vera-Ramirez MCharlet MParry GJ Recurrent aseptic meningitis compli-cating intravenous immunoglobulin therapy for chronic

inflammatory demyelinating polyradiculoneuropathy. *Neurology.* 1992;421636- 1637.

[9] Watson JDGGibson JJoshua DEKronenberg H Aseptic meningitis associated with high dose intravenous immunoglobulin therapy. *J Neurol Neurosurg Psychiatry.* 1991;54275- 276.

[10] Emmons CSmith JFlanigan M Cerebrospinal fluid inflammation during OKT3 therapy. *Lancet.* 1986;2510- 511.

[11] Roden JKlintmalm GBGHusberg BSNery JOlson LM Cerebrospinal fluid inflammation during OKT3 therapy. *Lancet.* 1987;2272.

[12] Thistlethwaite JRGaber AOHaag BW et al. OKT3 treatment of steroid-resistant renal allograft rejection. *Transplantation.* 1987;43176- 184.

[13] Rodríguez SC, Olguín AM, Miralles CP, Viladrich PF. Characteristics of me-ningitis caused by Ibuprofen: report of 2 cases with recurrent episodes and review of the literature. *Medicine (Baltimore).* 2006;85:214-220.

[14] Moris G, Garcia-Monco JC. The challenge of drug-induced aseptic meningitis. *Arch Intern Med.* 1999;159:1185-1194.

[15] Connolly KJ, Hammer SM. The acute aseptic meningitis syndrome. *Infect Dis Clin North Am.*1990;4:599-622.

[16] Cunha BA. The diagnostic usefulness of cerebrospinal fluid lactic acid levels in central nervous system infections. *Clin Infect Dis.* 2004;39:1260-1261.

[17] Morgan A, Clark D. CNS adverse effects of nonsteroidal antiinflammatory drugs: therapeutic implications. *CNS Drugs* 1998; 9: 281-90.

[18] Ruppert GB, Barth WF. Ibuprofen hypersensitivity in systemic lupus erythe-matosus. *South Med J* 1981; 74: 241-3.

[19] Durback MA, Freeman J, Schumacher VR. Recurrent ibuprofen-induced aseptic meningitis: third episode after only 200 mg generic ibuprofen. *Arthritis Rheum* 1988; 31.

[20] Katona BG, Wigley FM, Walters JK, et al. Aseptic meningitis from over-the-counter ibuprofen [letter]. *Lancet* 1988; I: 59.

[21] Lawson JM, Grady MJ. Ibuprofen-induced aseptic meningitis in a previously healthy patient. *West J Med* 1985; 143: 386-7.

[22] Gilbert GJEichenbaum HW Ibuprofen-induced meningitis in an elderly patient with systemic lupus erythematosus. *South Med J.* 1989;82514- 515.

[23] Ewert BH Resident article: ibuprofen-associated meningitis in a woman with only serologic evidence of a reumathologic disorder. *Am J Med Sci.* 1989;297326- 327.

[24] Grimm AMWolf JE Aseptic meningitis associated with nonprescription ibu-profen use. *DICP.* 1989;23712.

[25] Chez MSila CARansohoff RMLongworth DLWeida C Ibuprofen-induced meningitis: detection of intrathecal IgG synthesis and immune comple-xes. *Neurology.* 1989;391578- 1580.

[26] Widener HLLittman BL Ibuprofen-induced meningitis in systemic lupus erythematosus. *JAMA.* 1978;2391062- 1064.

[27] Weksler BB, Lehany AM. Naproxen-induced recurrent aseptic meningitis. *DICP* 1991; 25: 1183-4.

[28] Sylvia LM, Forlenza SW, Brocavich JM. Aseptic meningitis associated with naproxen. *DICP* 1988; 22: 399-401.

[29] Seaton RA, France AJ. Recurrent aseptic meningitis following non-steroidal anti-inflammatory drugs – a reminder. *Postgrad Med J* 1999; 75 (890): 771-2.

[30] Ballas ZK, Donata ST. Sulindac-induced aseptic meningitis. *Arch Intern Med* 1982; 142: 165-6.

[31] Lorino GD, Hardin JG. Sulindac-induced meningitis in mixed connective tissue disease. *South Med J* 1983; 76: 1185-7.

[32] Greenberg GN. Recurrent sulindac-induced aseptic meningitis in a patient tolerant to other nonsteroidal anti-inflammatory drugs. *South Med J* 1988; 81: 1463-4.

[33] von Reyn CF. Recurrent asepticmeningitis due to sulindac. *Ann Intern Med* 1983; 99: 343-4.

[34] Yasuda Y, Akiguchi I, KameyamaM. Sulindac-induced aseptic meningitis in mixed connective tissue disease. *Clin Neurol Neurosurg* 1989; 91 (3): 257-60.

[35] O'BrienWM. Adverse reactions to nonsteroidal anti-inflammatory drugs: dicl-ofenac as compared with other nonsteroidal anti-inflammatory drugs. *Am J Med* 1986; 80: 70-80.

[36] Joffe AM, Farley JD, Linden D, Goldsand G. Trimethoprim-sulfamethoxazole associated aseptic meningitis: case reports and review of the literature. *Am J Med.* 1989;87332- 338.

[37] Derbes SJ Trimethoprim-induced aseptic meningitis. *JAMA.* 1984;2522865- 2866.

[38] Creel GBHuitt M Cephalosporin-induced recurrent aseptic meningitis. *Ann Neurol.* 1995;37815- 817.

[39] Gordon MFAllon MCoyle PK Drug-induced meningitis. *Neurology.* 1990;40163- 164.

[40] Hedlund J, Aurelius E, Andersson J. Recurrent encephalitis due to trimethoprim intake. *Scand J Infect Dis* 1990; 22 (1): 109-12.

[41] Escalante A, Stimmler MM. Trimethoprim-sulfamethoxasole induced meningitis in systemic lupus erythematosus. *J Rheumatol* 1992; 19 (5): 800-2.

[42] Pashankar D, McArdle M, Robinson A. Co-trimoxazole induced aseptic me-ningitis. *Arch Dis Child* 1995; 73 (3): 257-8.

[43] Gilroy N, Gottlieb T, Spring P, et al. Trimethoprim-induced aseptic meningitis [letter]. *Lancet* 1997; 350 (9071): 112.

[44] de la Monte SM, Hutchins GM, Gupta PK. Aseptic meningitis, trimethoprim, and Sjogren's syndrome [letter]. *JAMA* 1985; 253 (15): 2192.

[45] Kremer I, Ritz R, Brunner F. Aseptic meningitis as an adverse effect of co-trimoxazole [letter]. *N Engl J Med* 1983; 308 (24): 1481.

[46] Biosca M, de la Figuera M, Garcia-Bragado F, et al. Aseptic meningitis due to trimethoprim-sulfamethoxazole. *J Neurol Neurosurg Psychiatry* 1986; 49: 332-3.

[47] Creel GB, Hurtt M. Cephalosporin-induced recurrent aseptic meningitis. *Ann Neurol* 1995; 37 (6): 815-7.

[48] Farmer L, Echlin FA, Loughlin WC, et al. Pachymeningitis apparently due to penicillin hypersensitivity. *Ann Intern Med* 1960; 52: 910-4.

[49] Garagusi VF, Neefe LI, Mann O. Acute meningoencephalitis associated with isoniazid administration. *JAMA* 1976; 235: 1141-2.

[50] Asperilla MO, Smego RA. Eosinophilic meningitis associated with ciprofloxa-cin. *Am J Med* 1989; 87: 589-90.

[51] Corson AP, Chretien JH. Metronidazole-associated aseptic meningitis [letter]. *Clin Infect Dis* 1994; 19 (5): 974.

[52] Buckley RM, Watters W, MacGregor RR. Persistent meningeal inflammation associated with intrathecal gentamicin. *Am J Med Sci* 1977; 274 (2): 207-9.

[53] Watson JD, Gibson J, Joshua DE, et al. Aseptic meningitis associated with high dose intravenous immunoglobulin therapy. *J Neurol Neurosurg Psychiatry* 1991; 54 (3): 275-6.

[54] Kato E, Shindo S, Eto Y, et al. Administration of immune globulin associated with aseptic meningitis. *JAMA* 1988; 259 (22): 3269-71.

[55]   Ellis RJ, Swendson MR, Bajorek J. Aseptic meningitis as a complication of intravenous immunoglobulin therapy for myasthenia gravis. *Muscle Nerve* 1994; 17: 683-4.

[56]   Vera-ramirez M, Charlet M, Parry GJ. Recurrent aseptic meningitis compli-cating intravenous immunoglobulin therapy for chronic inflammatory demyelinating polyradiculoneuropathy. *Neurology* 1992; 42: 1636-7.

[57]   Schiavotto C, RuggeriM, Rodeghiero F. Adverse reactions after high-dose in-travenous immunoglobulin: incidence in 83 patients treated for idiopathic thrombocytopenic purpura (ITP) and review of the literature. *Haematologica* 1993; 78 (6 Suppl. 2): 35-40.

[58]   Sekul EA, Cupler EJ, Dalakas MC. Aseptic meningitis associated with high-dose intravenous immunoglobulin therapy: frequency and risk factors. *Ann Intern Med* 1994; 121 (4): 259-62.

[59]   Dalakas MC Aseptic meningitis and intravenous inmmunoglobulin therapy. *Ann Intern Med.* 1995;122316- 317.

[60]   Jolles S, Hill H. Management of aseptic meningitis secondary to intravenous immunoglobulin [letter, comment]. *BMJ* 1998; 316 (7135): 936.

[61]   Nydegger UE. Safety and side effects of i.v. immunoglobulin therapy. *Clin Exp Rheumatol* 1996; 14 Suppl. 15: 53S-57S.

[62]   Ansell G. Toxicity of opaque media used in X-ray diagnosis. In: Davies DM, editor. *Toxicity of opaque media used in X-ray diagnosis: textbook of adverse drug reactions.* New York: Oxford University Press, 1981: 576-95.

[63]   DiMario Jr FJ. Aseptic meningitis secondary to metrizamide lumbar myelo-graphy in a 41/2-month-old infant. *Pediatrics* 1985; 76 (2): 259-62.

[64]   Forster G, Sacks S, Christott N. Aseptic meningitis as a complication of scinticysternography utilizing 111indium-DTPA. *Clin Neurol Neurosurg* 1975; 78 (4): 289-92.

[65]   Kalyanaraman K. Eosinophilic meningitis after repeated Iophendylate injection myelography [letter]. *Arch Neurol 1980*; 37 (9): 602.

[66]   Eustace S, Buff B. Magnetic resonance imaging in drug-induced meningitis. *Can Assoc Radiol J* 1994; 45 (6): 463-5.

[67]   Goldstein NP, McKenzie BF, McGukin WF, et al. Experimental intrathecal ad-ministration of methylprednisolone acetate in multiple sclerosis. *Trans Am Neurol Assoc* 1970; 95: 243-4.

[68] Sehgal AD, Tweed DC, Gardner WJ, et al. Laboratory studies after intrathecal corticosteroids. *Arch Neurol* 1963; 9: 74-8.

[69] PlumbVJ, DismukesWE. Chemical meningitis related to itrathecal corticosteroid therapy. *South Med J* 1977; 70: 1241-3.

[70] Gutknecht DR. Chemical meningitis following epidural injections of corticosteroids [letter]. *Am J Med* 1987; 82: 570.

[71] Stratton I. Dangers of intrathecal hydrocortisone sodium succinate [letter]. *Med J Aust* 1975; 18 (2): 650.

[72] Mott MG, Stevenson P, Wood CB. Methotrexate meningitis. *Lancet* 1972; II (7778): 656.

[73] Lazarus HM, Herzig RH, Herzig GP, et al. Central nervous system toxicity of high-dose systemic cytosine arabinoside. *Cancer* 1981; 48: 2577-82.

[74] Wolff L, Zighelboim J, Gale RP. Paraplegia following intrathecal cytosine ara-binoside. *Cancer* 1979; 43: 83-5.

[75] Gibbons RB. Chemical meningitis following spinal anesthesia. *JAMA* 1960; 29: 94-101.

[76] Dunkley B, Lewis TT. Meningeal reaction to starch powder in the cerebrospi-nal fluid. *BMJ* 1977; 2: 1391-2.

[77] Miller E, Goldacre M, Pugh S, et al. Risk of aseptic meningitis after measles, mumps, and rubella vaccine in UK children. *Lancet* 1993; 341 (8851): 979-82.

[78] Nalin DR. Mumps vaccine meningitis: which strain? [letter, comment]. *Lancet* 1989; 2 (8676): 1396.

[79] Cizman M, Mozetic M, Radescek-Rakar R, et al. Aseptic meningitis after vac-cination against measles and mumps. *Pediatr Infect Dis J* 1989; 8 (5): 302-8.

[80] Tesovic G, Begovac J, Bace A. Aseptic meningitis after measles, mumps, and rubella vaccine [letter, comment]. *Lancet* 1993; 341 (8859): 1541.

[81] Heinzlef O, Moguilewski A, Roullet E. Acute aseptic meningitis after hepatitis B vaccination [letter]. *Presse Med* 1997; 26 (7): 328.

[82] Hemet C, Chassagne P, Levade MH, et al. Aseptic meningitis secondary to carbamazepine treatment of manic-depressive illness [letter]. *Am J Psychiatry* 1994; 151 (9): 1393.

[83] Dang CT, Riley DK. Aseptic meningitis secondary to carbamazepine therapy. *Clin Infect Dis* 1996; 22 (4): 729-30.

[84] Sergent JS, Lockshin M. Azathioprine-induced meningitis in systemic lupus erythematosus [letter]. *JAMA* 1978; 240 (6): 529.

[85] Green MA, Abraham MN, Horn AJ, Yates TE, Egbert M, Sharma A. Lamotrigi-ne-induced aseptic meningitis: a case report. *Int Clin Psychopharmacol.* 2009 May;24(3):159-61.

[86] Martin MA, Massanari RM, Nghiem DD, et al. Nosocomial aseptic meningitis associated with administration of OKT3. *JAMA* 1988; 259 (13): 2002-5.

[87] Kreis H. Adverse events associated with OKT3 immunosuppression in the prevention or treatment of allograft-rejection. *Clin Transplant* 1993; 7: 431-46.

[88] Falloon J, Owen C, Kovacs, et al. MK-639 (MerckHIVprotease inhibitor) with in-terleukin-2 (IL2) in HIV. *Program and abstracts of the 35th Interscience Con-ference on Antimicrobial Agents and Chemotherapy* (ICAAC); 1995 Sep 17-20, 1995, San Francisco (CA): American Society of Microbiology (electronic version), Abstracts-On-Disk (R), Marathon Multimedia, Northfield, MN, 1995.

[89] Durand JM, Suchet L, Morange S, et al. Ranitidine and aseptic meningitis. *BMJ* 1996; 312 (7035): 886.

[90] Merrin P, Williams IA. Meningitis associated with sulphasalazine in a patient with Sjogren's syndrome and polyarthritis. *Ann Rheum Dis* 1991; 50 (9): 645-6.

[91] Alloway JA, Mitchell SR. Sulfasalazine neurotoxicity: a report of aseptic me-ningitis and a review of the literature [letter]. *J Rheumatol* 1993; 20 (2): 409-11.

[92] Herlihy TE. Phenazopyridine and aseptic meningitis. *Ann Intern Med* 1987; 106 (1): 172-3.

[93] Bodokh I, Lacour JP, Costa I, et al. Aseptic meningitis during pyrazinamide therapy for lupus erythematosus. *Presse Med* 1993; 22 (12): 595-6.

In: Meningitis
Editor: Anthony L. Shrader

ISBN: 978-1-63117-136-9
© 2014 Nova Science Publishers, Inc.

Chapter 2

Bacterial Meningitis: Procalcitonin Levels, Initial Signs and Symptoms, and Complications in a Reanimation Unit

*Carmen Mateo Revilla[*1], Tomas Rodríguez Delgado[1], Juan José Gómez Sainz[1] and Luciano Aguilera Celorrio[2]*

[1]Anesthesia and Reanimation Unit, Basurto Hospital. Bilbao, Spain
[2]Doctor of Pharmacology at the University of the Basque Country,
Department of Surgery, Radiology and Physical Medicine,
Faculty of Medicine University of the Basque Country,
Director of the Anesthesia and reanimation unit,
Basurto hospital, Bilbao, Spain

[*] Email: carmen.mateorevilla@osakidetza.net

Abstract

We conducted a retrospective review of patients hospitalized in our reanimation unit during a period of time of 53 months (from January 2009 to May 2013), with the diagnosis of bacterial meningitis. The main goal of this chapter was to analyze the symptoms, APACHE II, SOFA and procalcitonin levels of these patients on admission and their evolution during their stay in the unit.

We evaluated 25 patients. Two of them were diagnosed nosocomial meningitis, and the rest community-acquired meningitis.

5 patients (20 %) had smaller procalcitonin levels than 0.5 ng / dl on admission and did not increase during their stay in our unit.

The most common signs and symptoms were (from most to least): fever (84 %), altered level of consciousness (60 %), headache (48 %), neck stiffness (44 %), vomiting (40 %), ear pain (32 %), back pain (12%), neck pain (8 %), petechiae (8 %), photophobia (4 %), and seizure (4%).

The most frequently isolated bacterium was pneumococcus in 14 patients (56 %), followed by Neisseria meningitidis in 3 patients (13.6 %) and Streptococcus pyogenes in 2 patients (9.1 %). Staphylococuss aureus, Escherichia coli, Haemophilus, Listeria and Bacteroides fragilis were each found in one patient.

Most frequent complications were: requirement of mechanical ventilation (44 %), coma (40 %) and septic shock in (32 %). Followed by rhabdomyolysis (20 %), seizures (12 %), pneumonia (8 %), neurological deficit in (8 %), urinary tract infection (4 %) and brain death (4 %).

Mortality rate in patients was 12 %. The patients who died had an APACHE II mean of 23 and SOFA of 7.67 on admission, compared with those discharged from the RU who had an APACHE II mean of 14.7 and SOFA 3.25.

Introduction

Despite advances in antibiotherapy, bacterial meningitis carries a high morbidity and mortality that is why an early diagnosis and treatment are essential for survival. [1] In order to start an antibiotic treatment, blood tests, blood cultures and lumbar puncture (LP) when possible should be done quickly to all patients with suspected meningitis. The cerebrospinal fluid (CSF) cultivation is not immediate. Decisions must be taken in many cases based on the CSF biochemistry, which sometimes does not clearly differentiate the bacterial meningitis. [2] Therefore we also rely on other parameters like procalcitonin (PCT), which predicts high probability of bacterial infection.

There are not many published works [3-6] in medical literature with PCT as an indicator of bacterial meningitis and in most of them high levels of PCT were found in adults and children. However Villalon et al. [4] described low levels of PCT in 2 out of 23 patients with meningitis, even though they were previously treated with antibiotics. Schwarz et al. [5] also found levels below 0.5µg / L of PCT in 5 patients.

Patients and Methods

We conducted a retrospective review of patients admitted during 53 months (from January 2009 to May 2013) in the reanimation unit (RU) of the Basurto University Hospital with the diagnosis of bacterial meningitis. Patients' age was over 16 years old.

We wrote down daily in our database, all data collected from each patient. (Table I)

Table I. Database

I.	Join date.
II.	Personal history.
III.	Diagnoses and procedures performed.
IV.	Analytics on admission.
V.	SOFA, APACHE II.
VI.	Mechanical ventilation, inotropic drugs, infections and other complications.
VII.	Clinical status at end of treatment.
VIII.	Length of stay in the RU.

We obtained 25 adult patients from the database with the diagnosis of bacterial meningitis. 12 of them were women and 13 men between 16 and 89 years old. Only 2 of them had nosocomial meningitis: one after a lumbar arthrodesis postoperative (caused by spinal stenosis) and the other one was a patient with pericarditis.

We analyzed as well the analytics and reports of admission and discharge from the RU of these 25 patients.

The patients´ admission in the RU was decided by the doctor on duty, according to our protocol. (Table II)

Blood cultures were taken before antibiotic treatment from all patients with suspected meningitis. In most cases a computed tomography (CT scan) was made before the lumbar puncture.

Table II. Criteria for admission to the RU

I.	> 60 years.	
II.	Glasgow Coma Scale < 10.	
III.	APACHE II > 13.	
IV.	Septic Shock.	
V.	Leukopenia.	
VI.	Petechiae.	
VII.	Isolation of pneumococcus or Gram- bacilli in blood cultures or CSF.	
VIII.	Renal failure, mechanical ventilation, seizures, severe agitation.	

If the patient had a rapid neurological deterioration; we should perform (according to our protocol) a CT scan before a lumbar puncture. We would do the same procedure on patients immunocompromised or presenting otitis, sinusitis and dental abscess since they could carry complications such as: brain abscesses, stroke or focal infection.

The most worrisome contraindication to lumbar puncture is the suspicion of increased intracranial pressure (ICP) due to a cerebral mass lesion. Performing a lumbar puncture in these patients may lead to either transtentorial or uncal herniation and acute neurological deterioration. Lumbar puncture should not be performed in case of a suspected increase of the intracranial pressure. However, requesting the CT scan worsens the prognosis of patients with meningitis. It could as well delay the beginning of antibiotic treatment, leading to an increase of mortality. [7]

The lumbar puncture was delayed 30 minutes in patients with seizures, since they are associated with a transient increase of the intracranial pressure due to an increase up to 600% of the cerebral blood flow. Antibiotherapy was administered to patients who were suspected to have meningitis (and lumbar puncture was delayed by the need of performing a CT scan) after performing blood cultures. Patients were diagnosed with meningitis if:

- There was isolation of bacteria in CSF.
- Bacteria were isolated in blood with clinical suspicion and laboratory data suggesting meningitis.
- We had clinical suspicion and laboratory data of bacterial meningitis but there was no isolation of bacteria in neither blood culture nor CSF.

Aspect, pressure, glucose, proteins and cells of the CSF were analyzed in order to make a diagnosis of bacterial meningitis and it was determined by the following alterations in the CSF: increase in the pressure, large number of polymorphonuclear leukocytes, high proteins and low glucose (ratio CSF/plasma <0.3). Biochemical analysis of the CSF was positive for bacterial meningitis in all 25 patients.

The CSF culture isolated bacteria in 22 cases (88%). In 2 of the 3 remaining cases (8%), bacteria were isolated in blood cultures. In one case (4%) the CSF of the patient suggested bacterial meningitis verifying the subdural empyema at autopsy.

Treatment of patients based upon our protocols was always done after the extraction of blood cultures and in most cases, after performing the lumbar puncture. If patients were under 50 years old, community-acquired meningitis was treated with vancomycin and 3rd generation cephalosporin. If patients were older than 50 years, alcoholic or chronically ill, ampicillin was added. One case of neurosurgical postoperative was treated with meropenem and vancomycin. For all of them was given corticotherapy along with antibiotic treatment.

We used the following variables to develop this chapter:

- Month of admission, year, age, gender and length of stay in the reanimation unit.

- Clinical symptoms.
- APACHE II and SOFA scores on admission.
- CSF culture and blood cultures.
- Markers of infection such as PCT and C-reactive protein (CRP) on admission.
- Complications during the patients´ stay in the unit such as: mechanical ventilation, coma, septic shock, rhabdomyolysis, seizures, renal failure, neurological deficit, pneumonia, urinary tract infection and brain death.

Variables Description

SOFA *Sequential Organ Failure Assessment* was established in 1994 in a consensus meeting of the European Society of Intensive Care Medicine and was further revised in 1996. [8] It was created to measure organ dysfunction of 6 organs (pulmonary, hematologic, hepatic, cardiovascular, renal and central nervous system) and not to predict mortality, although its obvious relationship with mortality has been reflected in many studies. [9-10]

APACHE II *Acute Physiology and Chronic Health Evaluation II* was introduced in 1985 by Knaus and his collaborators. It is the most universally used system. It identifies clinical and physiological variables during the first 24 hours of the patients´ stay in the intensive care unit (ICU) and their history. It predicts probability of death depending on the score. According to Knaus, mortality expected in patients with APACHE $\geq$ 30 was 71.25 % - 82.5 %. Patients with APACHE 20-29 mortality expected was 40 % - 52.5 %. For those with APACHE 10-19 mortality expected was 12.5 %-a 22.5 %; and patients with APACHE 0-9 mortality expected was 3.75 % - 6.25 %. [11-13]

The Glasgow Coma Scale (GCS) was first introduced in the 1970s to provide a simple and reliable method of recording the level of consciousness of patients and monitoring change. In essence, the GCS was developed to standardize the reporting of neurologic findings and to provide an objective measure of the level of function of comatose patients. Originally developed as a series of descriptions of eye opening, motor and verbal responses, the addition of values or scores for each aspect of the GCS were added later. [14-15]

Patients from our unit with Glasgow Coma scale less than or equal to 8 were considered unable to protect their airway, requiring orotracheal intubation and mechanical ventilation.

Septic Shock, systemic illness caused by microbial invasion of normally sterile parts of the body is referred to as sepsis. When accompanied by evidence of hypoperfusion or dysfunction of at least one organ system, this becomes severe sepsis. Finally, where severe sepsis is accompanied by hypotension or need for vasopressors, despite adequate fluid resuscitation, the term septic shock applies. [16]

PCT is the prohormone of the calcitonin hormone. It is a biomarker for sepsis. PCT level rises rapidly in the first 6 to 8 hours after the onset of a bacterial infection. PCT concentrations in healthy individuals are below 0.05µg / L.

Usually, higher concentrations as 0.5 µg/L are interpreted as high values, suggesting septic syndrome. If PCT is less than 0.5 µg/L, bacterial sepsis is not likely but possible when the infection is localized or PCT levels are measured within 6 hours from the onset of infection.

It is not possible to exclude a systemic infection with PCT values between 0.5 and 2 µg/L. In these cases it is recommended to repeat the measurement 6 to 24 hours later to establish a specific diagnosis.

PCT levels above 2 µg / L indicate a high likelihood of bacterial sepsis with systemic consequences. Concentrations above 10 µg / L are found almost exclusively in patients with severe sepsis or septic shock. [17]

Results and Discussion

Between January 2009 and May 2013, twenty five patients diagnosed with bacterial meningitis were admitted in our Reanimation Unit. Thirteen of them were men (52%) and in most of the cases (92 %), the origin of the infection happened outside the hospital.

The median age of those patients was 58 years old, range 15 to 89 years old (mean = 59.8; standard deviation or SD = 16.8). The average stay in the hospital was 4.64 days (SD= 4.0; range 1 to 15 days). Three patients died during treatment. Therefore we are talking about a 12 % of mortality.

The result obtained from the APACHE II score was a median of 14, with a minimum of 3 points to a maximum of 35 (mean = 15.3; SD = 7.0). From the SOFA score was obtained a median of 3 (range / interval = 1-10; median = 3.8; SD − 2.4).

We analyze the possible association between the SOFA score and APACHE II score, with mortality and length of stay in the reanimation unit: 3

out of 25 patients with meningitis that died, had a SOFA score mean of 7.67 (SD = 2.51) on the day of admission in the unit. However, the other 20 patients (the ones who were discharged from the RU) had a mean of 3.25 (SD = 1.83). This difference (4.42; CI 95 %: 1.96 to 6.87) was statistically significant (p = 0.001) even when, due to the limited sample size (U Mann-Whitney; p = 0.012), nonparametric tests were applied. We can conclude that mortality and the SOFA score were related.

There is as well a positive relation between the SOFA score and the length of the patients´ stay in the reanimation unit (r = 0.48; p = 0.020). However and possibly as a result of the limited size of our sample, this association loses its statistical significance when analyzed using nonparametric statics (Spearman Rho = 0.356; p =0.095).

During the first 24 hours in the reanimation unit, APACHE II score was statistically associated with mortality. The patients who died had an APACHE II score mean of 23.00 compared with the 14.27 of those who were discharged from the unit (mean difference = 8.7; CI 95 %: 0.46 to 16.098; p = 0.039). However, in the nonparametric tests, this statistical association disappeared (U Mann-Whitney; p = 0.151).

There was no correlation between a higher value of APACHE II score and an increase in the length of the patients´ stay in the unit (r = 0.24; p = 0.24; Spearman Rho = 0.35; p = 0.09).

The median Glasgow Coma Scale of the patients upon arrival in the unit was 10, with a range between 3 and 15 (mean = 10.12; SD = 3.60). We could observe in our patients, that GCS was much related to mortality (the less GCS, the higher mortality rate). The GCS mean of the patients who died was 6.00 (SD = 3.00), and 10.68 (SD = 3.34) in those who were discharged, with a p = 0.031 of the t Student statistic.

We could notice on the patient´s clinical features on the day of admission in the RU, that the classic triad (fever, neck stiffness and altered consciousness) was notable in only 7 out of 25 patients (28.0 %). However just one of those three symptoms appeared in almost all of them (96.0 %). On a review conducted in 696 cases of community-acquired pneumonia, the triad was present in only 44 % of episodes [18].

We also discussed the signs and symptoms of patients separately. (Table III) Petechiae appeared in only 2 out of 3 patients with Neisseria meningitidis.

From the microbiological point of view, bacteria were isolated in CSF (22 patients – 88.0 %), in blood cultures (15 patients - 60.0 %), and in both (24 patients – 96.0 %).

Table III. Signs and symptoms of meningitis

Symptoms, signs	No. patients	%
Fever	21	84
Altered level of consciousness	15	60
Neck stiffness	11	44
Headache	12	48
Vomiting	10	40
Ear pain	8	32
Back pain	3	12
Neck pain	2	8
Petechiae	2	8
Cough	2	8
Seizure	1	4
Photophobia	1	4

With the exception of a patient diagnosed by his symptoms and the biochemistry of the CSF. The bacteria most commonly isolated in CSF and blood cultures were the pneumococcus (14 patients – 63.6 % of the total infections in CSF). Followed by Neisseria meningitidis (3 patients – 13.6 %) and Streptococcus pyogenes (2 patients - 9.1 %). A different kind of the following bacteria were isolated in each of the remaining five patients: Listeria, Escherichia coli, Haemophilus, Bacteroides fragilis and Staphylococcus aureus.

In all pneumococcal meningitis, pneumococcus was detected in the CSF. Only 8 out of 14 patients (57.1%) had positive blood cultures. Neisseria meningitidis was detected in all cases in the CSF. Only 1 out of 3 patients had positive blood cultures. Only blood cultures were positive in patients with Haemophilus and Bacteroide fragilis.

The main complications detected in our unit are described in the table. (Table IV)

Only 6 out of 14 patients (42.9 %) with pneumococcal meningitis needed mechanical ventilation. A Glasgow Coma Scale under eight was obtained in 6 patients (42.9 %). 5 patients suffered septic shock (35.7 %). 3 patients had seizures (21.4 %). 3 other patients had rhabdomyolysis (21.4 %). However the 3 patients with Neisseria meningitidis did not suffer any of these complications. Patients with pneumococcal meningitis had more frequent and severe complications, as it happened as well to those from other published articles. [18]

Table IV. Complications

Complications	No. patients	%
Mechanical Ventilation	11	44
Coma (G. C. S < 8)	10	40
Septic Shock	8	32
Rhabdomyolysis	5	20
Seizures	3	12
Renal failure	2	8
Neurological deficit	2	8
Pneumonia	2	8
Urinary Tract Infection	1	4
Brain death	1	4

Table V. CRP and PCT

patient	septic shock	CRP admision	PCT admision	maximum CRP	maximum PCT
1	yes	8,7 mg / dl	5,6 ng / ml	33 mg / dl	12,6 ng / ml
2	no	6,2 mg / dl	0,5 ng / ml	35 mg / dl	0,5 ng / ml
3	yes	4,8 mg / dl	6,3 ng / ml	27 mg / dl	6,3 ng / ml
4	yes	31 mg / dl	15,6 ng / ml	42 mg / dl	89 ng / ml
5	no	26 mg / dl	1,8 ng / ml	30 mg / dl	3 ng / ml
6	no	12 mg / dl	5 ng / ml	26 mg / dl	5 ng / ml
7	yes	18 mg / dl	1 ng / ml	28 mg / dl	1,3 ng / ml
8	yes	14 mg / dl	7,8 ng / ml	29 mg / dl	7,8 ng / ml
9	no	20 mg / dl	0,3 ng / ml	20 mg / dl	0,2 ng / ml
10	no	43 mg / dl	1,9 ng / ml	43 mg / dl	14,9 ng / ml
11	no	5 mg / dl	0,27 ng / ml	18,5 mg / dl	0,4 ng / ml
12	no	12 mg / dl	14 ng / ml	28,6 mg / dl	14 ng / ml
13	no	13 mg / dl	0,23 ng / ml	13 mg / dl	0,23 ng / ml
14	yes	57 mg / dl	11,9 ng / ml	71 mg / dl	17 ng / ml
15	no	28 mg / dl	1,98 ng / ml	35 mg / dl	2,96 ng / ml
16	no	20,4 mg / dl	0,9 ng / ml	20,4 mg / dl	0,9 ng / ml
17	no	20,1 mg / dl	0,11 ng / ml	20,1 mg / dl	0,11 ng / ml
18	no	38 mg / dl	36 ng / ml	39,4 mg / dl	36 ng / ml
19	no	22,8 mg / dl	29,7 ng / ml	34,2 mg / dl	36,5 ng / ml
20	yes	2,4 mg / dl	9,2 ng / ml	23 mg / dl	9,2 ng / ml
21	yes	0,26 mg / dl	0,12 ng / ml		
22	no	28 mg / dl	10 ng / ml	28 mg / dl	10 ng / ml
23	no	27 mg / dl	24 ng / ml	35 mg / dl	22 ng / ml
24	no	6 mg / dl	1,4 ng / ml	6 mg / dl	1,4 ng / ml
25	no	9 mg / dl	2,7 ng / ml	32 mg / dl	6 ng / ml

We could observe, when analyzing the acute-phase reactants in patients on admission to the RU and their maximum range, that the minimum CRP was 0.26 mg / dl up to a maximum of 57.0 mg / dl with a mean of 18.91 and a standard deviation of 13.75. The mean of the maximum CRP that patients had during their stay in the RU was 29.88 mg / dl (SD = 12.44 mg / dl) and a range between 6.0 – 71.0 mg / dl. (Table V).

The PCT mean on admission was 7.53 ng / ml (SD = 9.74) and range 0.11 to 36 ng / ml. 5 out of 25 patients (20.0 %) with a lower initial PCT of 0.5ng / ml, of which one patient developed shock. 7 patients (28.0 %) were admitted with a value of PCT among 0.5 to 2 ng / ml, and one of them developed complications with septic shock. The other 13 patients (52.0 %) were admitted with levels above 2 ng / dl, developing 6 of them shock.

The maximum PCT that patients had during their stay in the RU varied between a minimum value of 0.11 ng / ml and a maximum of 89 ng / ml, with a mean of 12.38 ng / ml (SD = 19.3). While in the unit, the PCT value in 4 patients did not exceed 0.49 ng / ml and 16 patients had PCT values greater than 2 ng / ml. Therefore the PCT in 3 patients with levels under 2 ng / ml on admission increased during their stay in the RU. (Table V)

Conclusion

We emphasize in this chapter that procalcitonin levels were in 5 patients (20%) less than 0.5 ng / ml on admission and did not increase during their stay in the RU. We were not able to obtain new levels from only one patient, who had septic shock, complications and brain death in less than 2 hours. Therefore we cannot exclude bacterial meningitis just because of low procalcitonin values. Equally important is to highlight that the Glasgow Coma Scale, APACHE II and SOFA on admission were associated with mortality in our patients, but none of them were related to the length of stay in the unit.

The classic triad of fever, neck stiffness, and altered consciousness was present in only 28% of the patients. However 96% of them had at least one of the three symptoms. The bacteria most frequently isolated were pneumococcus, and the most frequent complications were mechanical ventilation, coma and septic shock. However, those patients with Neisseria meningitidis had no complications. In conclusion patients with pneumococcal meningitis had more frequent and severe complications.

The study was limited by the small sample size. Therefore we should take the results cautiously, since it is retrospective and might be subject to bias.

References

[1]   Durand ML, Calderwood SB, Weber DJ, et al.: Acute bacterial meningitis in adults. A review of 493 episodes. *N. Engl. J. Med* 1993; 328: 21-28.

[2]   Aronin Sl, Peduzzi P, Quagliarello VJ: Community-acquired bacterial meningitis: risk stratification for adverse clinical outcome and effect of antibiotic timing. *Ann. Intern Med.* 1998; 129: 862-869.

[3]   Gendrel D, Raymond J, Assicot M, et al.: Measurement of procalcitonin leels in children with bacterial or viral meningitis. *Clin. Infect. Dis* 1997; 24: 1240-1242.

[4]   Viallon A, Zeni F, Lambert C, et al.: High sensitivity and specificity of serum procalcitonin levels in adults with bacterial meninigitis. *Clin. Infect Dis* 1999;28:1828-1832.

[5]   Schwarz S, Bertram M, Schwav S, et al.: Serum procalcionin levels in bacterial and abacterial meningitis. *Crit. Care Med* 2000; 28: 1828-1832.

[6]   M. Jereb, I. Muzlovic, S. Hojker, et al.: Predicitive value of serum and cerebrospinal fluid procalcitonin levels for the diagnosis of bacterial meningitis. *Infection* 2001; 29: 209-212.

[7]   Aronin SI, Peduzzi P, Quagliarello VJ. Community-acquired bacterial meningitis: risk stratification for adverse clinical outcome and effect of antibiotic timing. *Ann. Intern Med* 1998; 129: 862-869.

[8]   J. -L. Vincent, R. Moreno, J. Takala, et al. The SOFA (Sepsis-related Organ Failure Assessment) score to describe organ dysfunction/failure. *Inten Care Med* 1996; 22: 707-710.

[9]   Tran DD, Groeneveld ABJ, Vander Meulen J, et al. Age, chronic disease, sepsis, organ system failure, and mortality in a medical intensive care unit. *Crit. Care Med.*1990;18:474-479.

[10] Flavio Lopes Ferreira, Daliana Peres Bota, Annette Bross, et al. Serial Evaluation of the SOFA Score to Predict Outcome in Critically Ill Patients. *JAMA.* 2001; 286:1754-1758.

[11] Knaus WA, Zimmermann JE, Wagner DP, Draper EA, Lawrence DE. APACHE acute physiology and chronic health evaluation: A

physiologically based classification system. *Crit. Care Med* 1981; 9:591-597.

[12] Torres Bonafonte O. Pronósticos de los ancianos con enfermedades agudas. [tesis] Barcelona, Esaña, 2007. http://www.tdcat.cesca.es/tesis_uab/available/tdx

[13] Knaus WA, Drapper EA, Wagner DP, Zimmerman JE. APACHE II: a severityof disease classification system. *Crit. Care Med* 1985; 13: 818-29.

[14] Belinda J Gabbe, Peter A Cameron, Caroline F Finche. The status of the GasgowComa Scale. *Emergency Medicine* 2003; 15: 353-360.

[15] Lynne Moore, André Lavoie, Stéphanie Camden. Stadistical Validation of the Glasgow Coma Score. *J. Trauma* 2006; 60:1238-1244.

[16] Levy MM, Fink MP, Marshall JC, et al; SCCM/ESICM/ACCP/ATS/

[17] SIS: 2001 SCCM/ESICM/ACCP/ATS/SIS International Sepsis definitions Conference. *Crit Care Med* 2003; 31:1250–1256.

[18] Jerome Pugin, Michael Meisner, Alain Leon et al. Guía para el uso clínico de la procalcitonina. *Thermo Fisher Scientific*. 2011.

[19] Diederik van de Beek, M.D, Ph. D, et al. Clinical Features and Prognostic Factors in Adults whit Bacterial Meningitis. *N Engl. J. Med* 2004; 351:1849-59.

[20] Tauber MG, Moser B: Cytokines and chemokines in meningeal inflammation: biology and clinical implications. *Clin. Infet Dis* 1999; 28: 1 12.

In: Meningitis
Editor: Anthony L. Shrader

ISBN: 978-1-63117-136-9
© 2014 Nova Science Publishers, Inc.

Chapter 3

Tuberculous Meningitis: Update to an Old Disease

Fernando Alarcon[1,*] and Gonzalo Dueñas[2]
[1]MD Chief and Professor of Neurology Department,
Hospital Eugenio Espejo,
[2]MD Chief and Professor of Radiology Department,
Hospital Metropolitano, Quito, Ecuador

Abstract

Tuberculous meningitis (TBM) is the most severe form of tuberculosis. It is an important cause of death and disease among both children and adults throughout the world, especially in developing countries.

The prognosis is worse in patients with HIV, in patients with severe neurological impairment, in chronic cases, and in patients with a resistance to drugs. The disease's pathogenesis has not as yet been sufficiently understood. Despite progress achieved over the past decades, a diagnosis of TBM is difficult to reach in many cases, because clinical features are non-specific and because there is not as yet any diagnostic test sufficiently sensitive, targeted and timely. The percentage of patients where a definitive diagnosis of TBM can be established is low. TBM

[*] E-mail: falarcn2000@hotmail.com; Phones: 593 2 2221 202; 593 2 2503 296.

diagnosis and treatment are a challenge for neurologists. There is no characteristic medical profile, which makes it difficult to reach a diagnosis. Systemic symptoms of infection by Mycobacterium tuberculosis (MT), along with meningeal signs, may be suggestive of tuberculous meningitis. HIV infection does not change how tuberculous meningitis is presented.

In 80% of patients with TBM, symptoms appear at least one week before diagnosis. The time profile for the onset of TBM symptoms could be used to differentiate tuberculous meningitis from bacterial meningitis. The percentage of patients diagnosed as having tuberculous meningitis might increase if a molecular diagnostic test were applied. Hydrocephalus, pre-contrast hyperdensity-hyperintensity, basal leptomeningeal enhancement, obliteration of the basal cisterns, infarcts and tuberculomas are highly frequent findingson CT and MRI studies. The diagnostic approach is based on clinical criteria, laboratory test results, and image findings.

Treatment delay is strongest risk factor for death. TBM prognosis depends on the presence of HIV, resistance to drugs, abnormal CT or MRI, substantial rise of proteins in CSF and onset of antituberculosis treatment three days after admission to hospital. The need to reach a diagnosis and administer early antituberculosis treatment is very important to reduce death and improve the prognosis. Severity on presentation is prognostic for outcome. Recently, a consensus in terms of clinical diagnostic, laboratory and imaging criteria have been defined and recommended for tuberculous meningitis to be used in future clinical research.

Introduction

One third of the world's population is infected by the tubercle bacillus. Autopsy series have revealed that 5-15% of people exposed to MT develop symptomatic pulmonary disease, and 5-10% of these individuals eventually develop CNS disease, most commonly meningitis [1]. HIV infection is the most important risk factor for the development of tuberculosis and carries a poor prognosis [2]. TBM is the most severe form of tuberculosis and is a major cause of disease and death among adults and children. TBM primarily affecting brain and spinal cord meninges, along with the adjacent brain parenchyma, tends to predominate in children from developing countries and adults from developed countries. TBM characterized by slow progressive granulomatous inflammation of the basal meninges, leading to hydrocephalus, cerebral infarct or death if the patient is not treated, is the most frequent form of tuberculous CNS [1, 3]. Formation of tuberculomas and abscesses, as well as affectation of the spinal cord, may also occur [2, 3]. Although anti-tubercular treatment is increasingly effective, disease and death rates continue to be high, making CNS tuberculosis a potentially devastating disease. Clinical, laboratory and imaging findings are not very sensitive or specific for the diagnosis. Diagnosing CNS tuberculosis is, in many cases, a true challenge, owing to the wide variability of clinical signs and symptoms and the absence of sensitivity of acid-fast bacillus (AFB) smear and culture of CSF [4]. A better prognosis can be ensured when treatment starts early in the disease [5]. Recent reviews have been published on the pathogenesis, diagnosis and treatment of tuberculous meningitis, highlighting major controversial issues that have been overturned or mistakenly accepted in the literature [6]. One benefits of the unequivocal of bacillus Calmette Guerin (BCG) vaccination is protection against disseminated forms of childhood tuberculosis specially meningitis and military

Epidemiology

MT infection´s incidence has risen drastically over recent years, not only in areas that were traditionally endemic but also in areas where the incidence of tuberculosis has been declining [7]. Although the lungs are the primary target of the infection, many organs are potentially affected and it is estimated that 10% of immunocompetent patients who have tuberculosis develop the disease in the CNS. In United States, 5 to 9% of AIDS patients have pulmonary or extrapulmonary tuberculosis. Tuberculosis is the most common opportunistic infection in persons infected with HIV. The main way of contagion is person-to-person transmission by inhalation of droplet nuclei that have aerosolized. These nuclei have few bacilli, but it is estimated that only 1 to 10 organisms are need to trigger the infection. The tubercles consist of mononuclear cells surrounding a caseous necrotic center that can be formed in the lung as well as in secondary sites [7]. The exact incidence and prevalence of TBM in HIV-infected patients is not known. Tuberculosis is a major cause of death among HIV-infected persons [8]. In 2008, WHO estimated that there were 33.4 million HIV-infected cases. That same year, there were about 1.4 million new cases of tuberculosis among HIV-infected persons and tuberculosis accounted for 23% of AIDS-related deaths. Worldwide, 14 million people are currently co-infected with tuberculosis and HIV [8]. In

2007, the estimated number of tuberculosis prevalent cases amounted to 13.7 million (206 per 100,000 inhabitants). During this period, 9.27 million new cases of tuberculosis were recorded worldwide and about 1.3 million died. Out of the new cases of tuberculosis, 1.37 million were HIV-positive [9].

Pathogenesis

TBM is caused by MT bacteria which is a gram-positive, aerobic, non-spore-forming, non-motile, pleomorphic rod and a acid-fast bacterium. This organism is an obligate aerobic bacillus whose entire genome has been sequenced. The genome has provided a wealth of information that may result in much-needed improvements in treatment, diagnosis and prevention. The most widely used acid-fast smear method for MT is the Ziehl-Neelsen stain. The Lowenstein-Jensen medium is the most popular and widely available culture medium to grow MT [7, 9]. TBM, which usually occurs as an immediate or remote complication of a primary infection or can develop over the course of chronic tuberculosis, can appear without any other evidence of associated pulmonary or extrapulmonary tuberculosis. In children miliary tuberculosis is directly involved in the pathogenesis of TBM, whereas in adults it is part of a widely disseminated tubercular process [3]. MT transmission to a healthy person is primarily by airborne droplet nuclei. In the lungs, MT bacteria multiplies in alveolar macrophages. Within two to four weeks, through blood circulation, bacilli spread to extrapulmonary sites and produce small granulomas in the meninges and adjacent brain parenchyma [9]. These small granulomas, now called Rich foci, are found predominantly within the brain parenchyma, but also in the meninges and in the subpial or subependymal surface of the brain [6, 9]. Rich foci remain dormant for years. Meningitis occurs when mycobacteria contained within these lesions are released into the subarachnoid space, and that might happen even months or years after the initial bacteremia [3, 6, 9]. There is no precise knowledge about the exact trigger for the rupture of Rich foci. Brain trauma, HIV infection and other factors that reduce immunity are believed to play a role [3, 9].

The ability of MT to spread in the blood seems to be the key component of the pathogenesis of tuberculosis [6]; however, it is not well understood how the bacteria pass from the lungs to the blood stream. Pulmonary alveolar macrophages represents the most plausible vehicle to transport MT from alveolar lumen to blood. Bacillemia associated with miliary dissemination enhances the probability of the formation of a meningeal or subcortical tuberculous focus, which may eventually caseate and give rise to TBM [9]. The bacillus passes into the CNS traversing the blood brain barrier. Persons infected with Koch bacillus experience a period of usually asymptomatic lymphohematogenous spread, which is the interval between infection and development of the tissue's specific hypersensitivity, manifested by a positive tuberculin reaction. This pre-allergic bacillemic phase usually establishes metastatic foci of infection that can remain inactive or become active after variable periods of clinical latency, which may be immediate or may last for months or years. It is likely that, in some cases, the foci of chronic tuberculosis become sources of secondary hematogenous spread, usually abortive by themselves, but capable of establishing other metastatic foci, including some adjacent to the subarachnoid space with a potential subsequent rupture. This mechanism could explain episodes of intermittent fever and general discomfort that are not well understood and that appear in some patients weeks or months before the development of TBM. This intermittent chronic form of presentation is associated with severe symptoms and is potentially disabling in patients with TBM [3].

Some strains of MT are considered to be highly virulent and are capable of causing more disseminated and meningeal forms of tuberculosis than others [9]. The different genotypes of MT may appear with variable clinical manifestations in the host. It has been suggested that some genotypes of MT are associated with drug resistant tuberculosis and with a high incidence of co-infection with HIV [9]. MT is an intracellular pathogen that survives within the phagosome of host macrophages. Apoptosis of infected macrophages is an effective mechanism used by the host against tubercle bacilli. Virulent strains of MT have evolved several genetic mechanisms to subvert host immune responses, leading to their prolonged survival and growth in the host's macrophages [9].

Clinical Features of Tuberculous Meningitis

The clinical manifestations of TBM are related to the pathological changes seen in this neuro-infection [3]. TBM has a non-specific clinical history. Its clinical presentation may differ, depending on whether it is occurring in adults or in children and whether or not it is occurring in HIV-infected patients [7].

Typically, a prodromal period has been recognized between the second and fourth week, which may extend from a few days to several months before its presentation [3-5, 7].

In our series [3], we have been able to identify three presentations in the prodromal period: an acute prodromal period, with signs and symptoms lasting up to one week from the onset of symptoms, accounting for 13% of cases; a subacute presentation lasting two to eight weeks, accounting for 80% of cases; and a chronic presentation lasting for eight weeks up to various months, accounting for 7% of cases [3]. In patients with a subacute presentation, three stages can be recognized [3]. The first stage, of general manifestations, courses with nonspecific symptoms, general discomfort, hyporexia, fever, intermittent headache and muscle aches. In this stage, infants may show irritability, abdominal pain and tense fontanelle. Children show meningism and gastro-intestinal symptoms. Elderly patients may shows states of confusion and absence of fever. This stage would be associated with the presence and initial spread of the tubercle bacillus in the subarachnoid space. The second stage courses especially with meningeal manifestations, neck pain and nuchal rigidity, fever, especially at night, continuous and severe headache, irritability, vomiting, cranial nerve palsy, somnolence, delirium and impaired consciousness. This second stage could be associated with the allergic reaction to the tuberculoprotein, which leads to changes in the CSF and basal exudate development. The third stage courses especially with cerebral or spinal parenchymatous manifestations, characterized by the presentation of papilledema, impaired consciousness, focal neurological deficit secondary to ischemia or cerebral infarct and compromise of the brain stem and the spinal cord or its roots. In this stage, patients may present with paraparesis or paraplegia and movement disorders that include chorea, ballism, tremor, parkinsonism, dystonia and myoclonus. Chorea-ballistic movements prevail in children [3, 10]. The third stage is associated with the progressive increase of brain and spinal meningeal exudate, vasculitis and vascular occlusion, cerebral edema, pressure distortion secondary to tuberculomas, hydrocephalus and intracranial hypertension [3, 11, 12]. Convulsions are present in 10 to 15% of cases [3].

In the acute presentation before admission, symptoms last for less than one week and can manifest as acute meningitis, clinically undifferentiated from pyogenic meningitis with fever, headache, nuchal rigidity and impaired consciousness, also capable of simulating viral encephalitis with unilateral focal deficit, convulsions, impaired consciousness and intracranial hypertension, especially in children. It is highly unlikely that there might be myeloradiculitis or myelopathy. Children can present movements disorders such as chorea and dystonia. The acute presentation could be related to a sudden discharge of a large amount of tuberculoprotein in the CSF [3].

In the chronic presentation, there might be symptoms that last more than eight weeks before admission. The patient has intermittent fever, progressive headache, apathy, cognitive impairment, irritability, gait ataxia, urinary incontinence, impairment of cranial pairs, delirium, tremor, myoclonus or other abnormal movements, motor impairment, with progressive hemiparesis, tetraparesis or paraparesis affecting the sphincter and, in some cases, radicular impairment. Paraplegia may be caused by tuberculous radiculomyelitis, arachnoiditis, intramedullary tuberculomas, or syringomyelia. This form of presentation, which is especially frequent in adults, corresponds to a late chronic form of neurotuberculosis, associated with reinfection or resurgence of tuberculous infection in the CNS with abortive episodes of spreading of the bacillus or with intermittent discharges of bacilli or tuberculous antigens from a caseous juxtaependymal focus in the subarachnoid space. The CSF in these cases can show a moderate to severe rise in proteins, up to four grams, with the response depending on the immunological reaction of the host [3].

In our series of 310 patients with TBM, the average age was 34.5 years (range from 0.5 to 80 years). The most frequent signs and symptoms were asthenia in 79.6% of patients, irritability in 79.5%, fever in 79%, headache in 78.4%, meningeal signs in 77.5%, altered mental state in 70%, weight loss in 64.9%, motor deficit in 54.7%, cranial nerve compromise in 25.4%, papilledema in 20.3%, abnormal movements in 18.4%, convulsions in 13.9%, disconjugate eye movements in 9.7%, nystagmus in 4.2%, and aphasia in 3.2%. Sixth cranial nerve was the most commonly affected. Third and fourth cranial nerves were less frequently affected [13].

Atypical clinical manifestations in the elderly may contribute to delay the diagnosis of TBM. In both elderly patients and children, meningeal signs are less frequent. In a study carried out in patients over 50 years of age [14], the main clinical manifestations were fever in 81% of cases, abnormal mental state in 72%, headaches in 47%, vomiting in 34%, rigidity in 51%, altered sensory in 64%, crises in 28%, and focal neurological deficit in 24%. TBM may appear in elderly patients or HIV patients with subacute dementia, personality changes and frontal lobe signs. In pediatric patients, prevailing clinical manifestations are coma, rise in intracranial pressure, crises, and focal neurological deficit [9].

Movement disorders may occur over the course of TBM [10]. Dystonia, chorea-ballism, tremor, myoclonus and parkinsonism have been found in

16.6% of patients with TBM [10]. Movement disorders are more common in children and young people. Chorea is most frequent in children under five years of age. Tremor is the most frequent abnormal movement. In patients with movement disorders because of TBM, there is a poor correlation between the type, distribution or severity of the abnormal movement and CT or MRI findings. In 60%, hydrocephalus was found and in 46.6% cerebral infarcts were found. The patients showed a high frequency of ischemic or hemorrhagic infarcts and exceptionally deep hematomas affecting the internal capsule, mesencephalus, basal ganglia and diencephalus. Vasculitis of the vessels surrounded by the purulent exudate, has been proposed as the cause of cerebral infarcts in patients with TBM. Ischemic infarcts in the basal ganglia, including the subthalamic nucleus as a consequence of vasculitis, is a probable mechanism for patients who show movement disorders. Hydrocephalus may play an important role in some cases. Multifactor mechanisms have been deemed responsible for the abnormal involuntary movements (AIMs) in patients with TBM [10].

One third of patients with brain tuberculomas (32.6%) may be associated with movement disorders [11, 12]. In patients with tuberculomas, chorea, which predominates in children, is the most frequent abnormal movement. Chorea and dystonia are usually associated with deep tuberculomas and tremor with superficial lesions. In patients with movement disorders, multiple tuberculomas were found in 68.7% of cases, with a predominant deep supratentorial location. In patients with chorea, there is a high correlation between motor deficit and abnormal movements (87.5%), suggesting that when patients have this association and show, in the image, one or various expansive lesions, especially in the basal ganglia, it is highly probable that its etiology is tuberculosis. The pathogenesis of movement disorders caused by intracranial tuberculomas has been related with a pressure-distortion mechanism that can be aggravated by edema, ischemia, toxic bacterial factors and hydrocephalus [12].

Dastur and Udani [16, 17] were the first to describe children with disseminated tuberculosis a variant that they called "tuberculous encephalopathy." These children have a cerebral disorder characterized by rapid and diffuse cerebral affectation with convulsions, involuntary movements, stupor or coma with absence of significant meningeal signs. CSF may show pleocytosis and normal or discreetly augmented proteins. Dastur has argued that the pathogenesis of tuberculous encephalopathy may differ from TBM. In post-mortem studies, diffuse cerebral edema has been found along with demyelinization and sometimes traces of hemorrhage, which may be more typical of post-infectious allergic encephalomyelitis [16On brain CT , there is extensive hypodensity of brain white matter [3]. Improvement has been reported when steroids were administered [3, 18].

TBM with spinal affectation, which commonly appears as paraplegia, is a severe complication in TBM, occurring in 10 to 15% of cases [3, 14]. It may be caused by vertebral tuberculosis (Pott's disease), which can be associated with fusiform paravertebral abscesses or a gibbus [3, 14, 19, 20] or by radiculomyelitis, myelitis, arachnoiditis (Figure 1), intradural or extraspinal tuberculomas (Figures 2, 3, 4), abscesses (Figure 5) or spinal syrigomyelic cavities. Tuberculous radiculomyelopathy characteristically appears with subacute paraparesis, radicular pain, paresthesia, bladder disorder and slowly progressive muscular weakness [3, 9, 14]. Muscle weakness is a late complication. In the neurological exam, there is sensitivity deficit, hypotonus and usually absence of deep reflexes. The plantar response may be extensor. Aabsence of deep reflexes provides a poor functional prognosis. Protein content in CSF of patients with paraplegia increases, becoming very high in some cases. This increase is related to partial or complete spinal blockage and especially with a gradual and progressive immunoallergic response to the presence of the tuberculoprotein and its antigens [3].

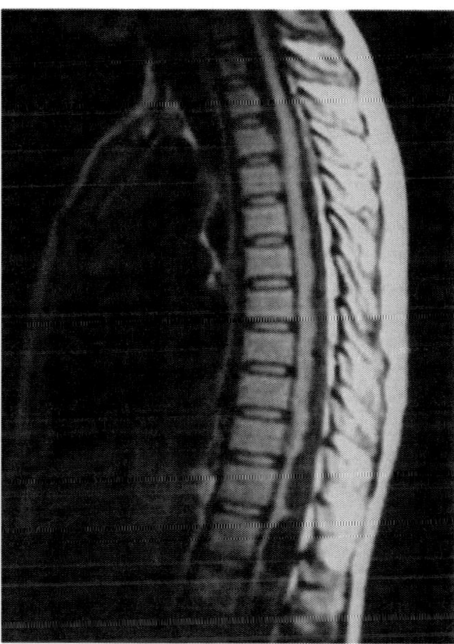

Figure 1. 42 year old patient. Sagital T1W spinal MRI shows myltiple leptomeningeal adherences and severe cord distortion.

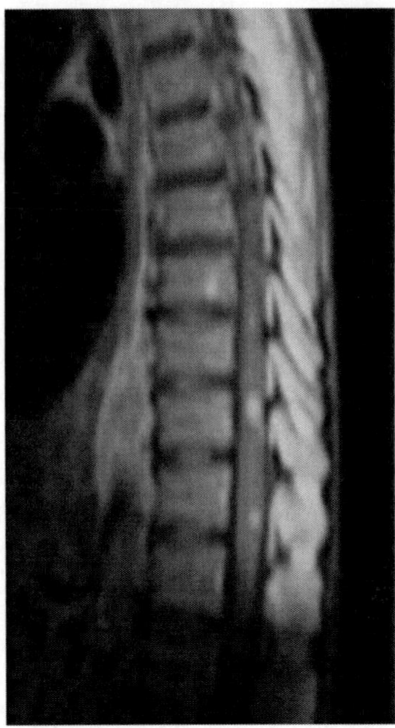

Figure 2. 17 year old patient. Sagital T1W spinal MRI with Gadolinium, shows multiple nodular enhansing spinal cord granulomas.

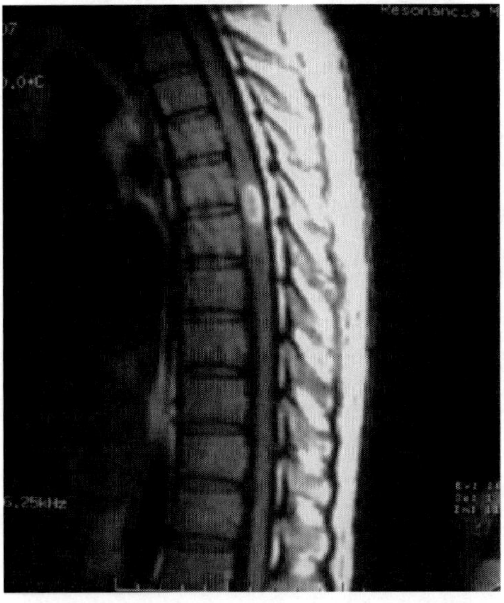

Figure 3. 56 year old patient. Sagital T1W spinal MRI with Gadolinium, shows a ring enhancing intramedulary granuloma.

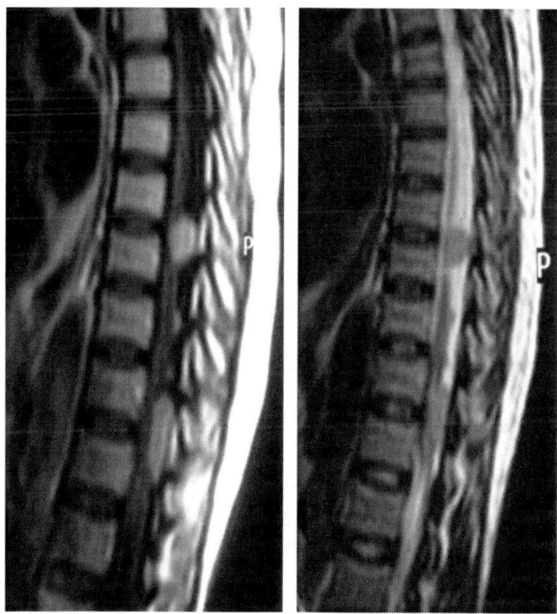

Figure 4. 35 year old patient. Sagital T1W with Gadolinium and T2W spinal MRI show intradural extramedulary granulomas and irregular leptomeningeal enhancement.

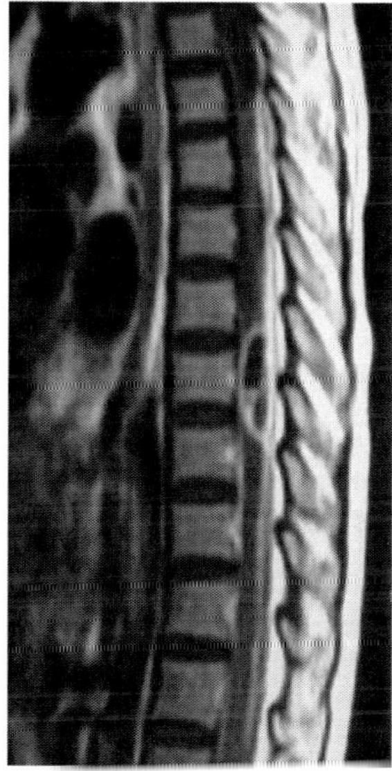

Figure 5. 28 year old patient. Sagital T1W spinal MRI with Gadolinium shows severe cord atrophy, meningeal enhancement and intradural extramedulary abscess.

Cranial nerves are affected either secondary to entrapment in thick basilar exudates or because of increased intracranial pressure [9]. The loss of sight in TBM may be related to optochiasmatic arachnoiditis, third ventricular compression of optic chiasm in patients with severe hydrocephalus, optic nerve granulomas or ethambutol toxicity [9].

Choroidal tubercles are infrequent funduscopic findings in patients with tuberculous meningitis. These lesions are considered pathognomonic for TBM. Choroidal tubercles are often associated with miliary tuberculosis [9]. In the examination of the fundus oculi, the choroidal tubercles are white, gray or yellow lesions with indistinct borders, surrounded by edema. Their size varies from 0.5 to 3 mm in diameter. Histologically choroidal tubercles represent caseated granulomas. Tubercle bacillus has been shown in choroidal tubercles [9].

The clinical features and CSF profiles of TBM are not modified by HIV infection. There are no significant differences between HIV-positive patients and HIV-negative patients, except for the presence of lymph node TB in 50% of HIV-positive patients, compared with 3% of HIV-negative patients. Some studies have reported a higher frequency of tuberculous intracerebral mass lesion in HIV-positive patients, mainly among intravenous drug users. Tuberculin skin tests are positive in about one third of patients, particularly in the early stages of HIV infection, while they develop an anergy with more advanced immunosuppresion [7].

Pathology

The pathological changes occurring in the cranium and spinal canal depend on various factors such as age, severity and duration of disease, patient's immunological condition or hypersensitivity, response to therapy and possibly the number and virulence of the bacillus, as well as the host's individual susceptibility to MT [3, 9, 20, 21]. Although the greatest impact of the disease takes place in the basal meninges and in the CSF, parenchymatous cerebral lesions, because of the direct extension of the inflammatory process or secondary to vascular changes, are found in most cases [3, 20]. Diffuse pathological changes affect the arachnoid membrane and the subarachnoid space. Pathological characteristics are meningeal inflammation, fibrogelatinous basal exudates, vasculitis of the arteries traversing the exudates and obstruction of the flow of CSF resulting in hydrocephalus, which can appear early or late and be of variable degree [3, 9]. The inflammatory exudate predominating in the interpeduncular and pontine cisterns, can extend to the sylvian fissure and other cisternsleading to blockage of the foramina of Luschka and Magendie. Leptomeningeal exudate progressively surrounds and strangles the brain stem and its arteries [3, 9].

The optic chiasm and the roots of other cranial nerves arising from the ventral aspect of the brainstem are usually entrapped in thick exudates [22, 23]. In addition, the exudate frequently covers the choroidal plexuses, the spinal canal, especially in the dorsal region and the ependyma of the ventricles, which has a granular aspect [3, 9]. Exudates can be seen surrounding the lower part of the spinal cord and cauda equina resulting in tuberculous radiculomyelopathy [3, 9]. Later the exudate can harden and calcify. In the meninges, choroidal plexuses, and ependyma, small nodes or tubercles more than 1 mm in diameter can be found [20, 21]

Microscopically, the exudate is predominantly composed of lymphocytes and plasma cells with caseous necrotic foci. When the process becomes chronic, the exudate becomes a fibrous mass compromising the cranial nerves and the spinal cord, conditioning secondary neurological manifestations [3, 20, 21]. TBM can result in the formation of tuberculomas consisting of caseous necrotic material, epithelioid cell granuloma and mononuclear cell infiltration [9].

Obstruction to the flow of CSF occurs in the posterior fossa, as the thick exudate blocks openings of fourth ventricles, the sylvian aqueduct or the opening of the tentorium, developing hydrocephalus as a consequence. In addition to the inflammatory exudate, obstruction of the sylvian aqueduct is related to ependymitis or the presence of tuberculomas [3, 20, 21]. Hydrocephalus may develop late as a result of arachnoid adherences [3]. The brain tissue underlying the tuberculous exudate shows various degrees of edema, perivascular infiltration and a microglial reaction collectively termed as "border zone encephalitis." A microscopic pathological feature of tuberculous meningitis is the formation of epithelioid cell granulomas with Langhans giant cells, lymphocytic infiltrates and caseous necrosis. Exudates are prominently present around the sylvian fissure, basal cysterns, brainstem and cerebellum [9].

The distribution of tubercles along the pial vessels, the occasional location of tuberculous lesions in an arterial territory and the presence of dense and fibrinocellular gelatinous leptomeningeal exudate contribute to the development of pathological findings related to stroke in TM [3, 24]. The leptomeningeal exudate initially located in the interpeduncular fossa is disseminated anteriorly and compromises the optic chiasm, anterior brain vessels and laterally the sylvian valley, surrounding the carotid artery, the middle cerebral artery and its penetrating branches [9].

Stroke occurs especially in the advanced and severe stage of the disease in 15 to 57% of cases [3, 10, 15, 18, 19, 24-26] and may be asymptomatic because it occurs more frequently in deep areas and not very manifest. Among children, infarcts of the basal ganglia and inner capsule have been associated with a poor prognosis [24-26]. Most strokes in TBM are multiple, bilateral and located in the basal ganglia, especially in the caudate, thalamus, anterior arm and knee of the inner capsule. These are attributed to compromise of the medial striatal perforating, thalamotuberal and thalamostriates arteries, which are immersed in the leptomeningeal exudate and can be compressed or displaced because of hydrocephalus. Cortical strokes can also occur because of the affectation of the proximal part of the middle (Figure 6), anterior and posterior cerebral arteries as well as the supraclinoid segments of the carotid arteries and the basilar artery. These findings have been documented in MRI, angiography and autopsy studies [3, 9, 24].

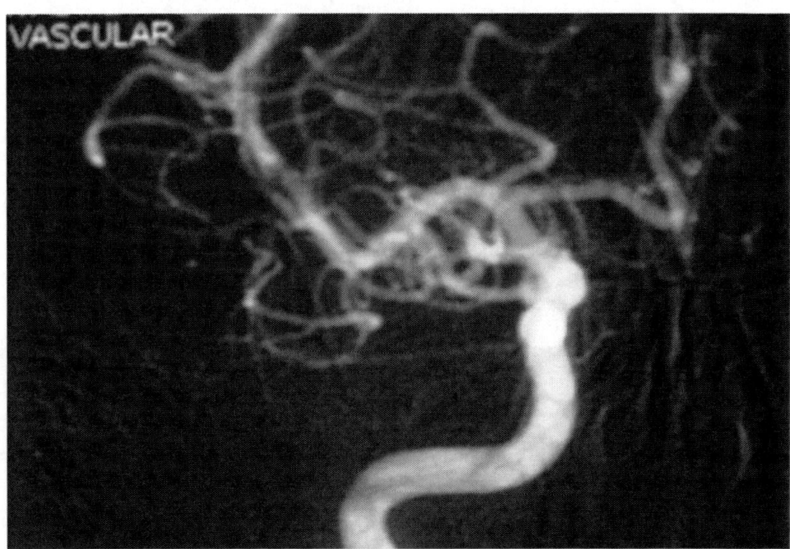

Figure 6. 27 year old patient. Cerebral angiography shows stenosis and intimal irregularity of the middle cerebral artery.

Changes in cerebral vessels are characterized by inflammation, spasm, constriction and eventually thrombosis of the vessels. Occlusion of cerebral arteries leads to infarction of the underlying tissue [9]. Arteritis is more common than infarct in autopsy studies. It has been suggested that cytokynes could play a role here, especially the tumoral necrosis factor, the endothelial vascular growth and the matrix of metaloproteinases in a damaged hematoencephalic barrier attracting leukocytes and liberating vasoactive autocoids. Without doubt, the prothrombic status may also contribute to the occlusion of the cerebral vessels in TBM. It has not been proven that the use of corticosteroids, antitubercular treatment and aspirin is able to reduce stroke frequency [24].

Diagnosis

CSF examination is crucial in determining the diagnosis of TBM. CSF pressure is increased in about 50% of cases. Changes in CSF help to differentiate it from other meningitis. The macroscopic aspect of the fluid may be clear or xanthochromic and may show a film or small lumps [3, 9]. Characteristically, CSF shows lymphocytic pleocytosis, higher concentration of proteins and low concentration of glucose. In 9 to 15% of patients, the first CSF does not show pleocytosis [3, 27]. Initial mononuclear pleocytosis may briefly change in direction to polymorphonuclear predominance when therapy starts, and this may be associated with clinical deterioration. This therapeutic paradox has been regarded as virtually pathognomonic for TBM [7]. In 81% of patients, moderate pleocytosis (between 5 and 500 cells/mm) can be found; below 100 cells/mm in 44% of cases. In 7% of patients there were between 500 and 1,500 cells/ml, and in 2% more than 1,500 cells/ml. In 72% of patients there was a predominance of lymphocytes [3], whereas 93% showed moderate elevation of proteins (between 30 and 500 mg/dl), 4% a substantial increase in proteins between 500 and 2,000 mg, and 2% a large increase of more than 2,000 mg/dl. In patients with a substantial increase in proteins, there is a high risk of affectation of the medulla. Glucose level declined to less than 45 mg/dl in 71% of cases, and in 10% it was less than 10 mg/dl [13]. The CSF changes reflect the individual reaction to the tuberculoprotein, which in some cases may not appear or is found to be diminished. In these cases, the white cell count or the level of proteins may be normal or slightly increased [3]. These small changes in CSF may be interpreted as excluding TM, which may be an error because it may be an authentic TBM or a serous form of TBM [3, 18], which is a self-confined meningeal disease affecting children with active pulmonary tuberculosis and which is considered to be caused by a reaction to a primary or post-primary focus in the CNS [3]. Increased proteins in CSF are almost constant. Nevertheless, the gold standard for diagnosis is the demonstration of MT bacilli in CSF. Unfortunately smear for acid-fast bacillus is positive only in 5-30% of patients in the first sample, with increasing possibly in the following samples [3, 9, 21, 28]. Culture of MT from CSF is not always positive and it takes several weeks for a positive result. Conventional CSF culture in a Lowenstein-Jensen medium is positive in 45-90% of cases [3, 9, 21], but results are only available after 6-8 weeks of incubation. In HIV-associated TM, 69% positivity in smears and 87.9% positivity in bacterial culture have been demonstrated [9, 29, 30]. To increase the positivity of the smear examination, it is necessary to collect 10 cc of CSF and centrifuge it at high speeds. The smear must be examined for at least 30 minutes [3, 27, 31]. A combination of both microscopic examination and culture of several samples of CSF (sometimes as many as four are needed) provides the highest yield of positivity [3, 7].

Because of the lower percentage of identification of MT in the first CSF sample and its slow growth in the culture mediums, fast tests have been developed as early diagnostic help methods in TBM. Biochemical test as Adenosine Deaminase (ADA) has been of interest for many years in TB diagnosis [3, 15].

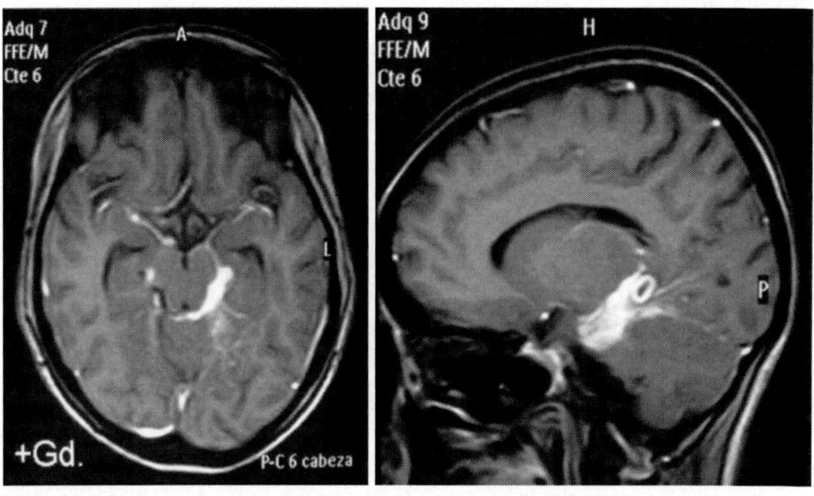

Figure 7. 17 year old patient. Axial and sagital T1W brain MRI with Gadolinium shows leptomeningeal enhancement around the mesencephalon and a supratentorial granuloma.

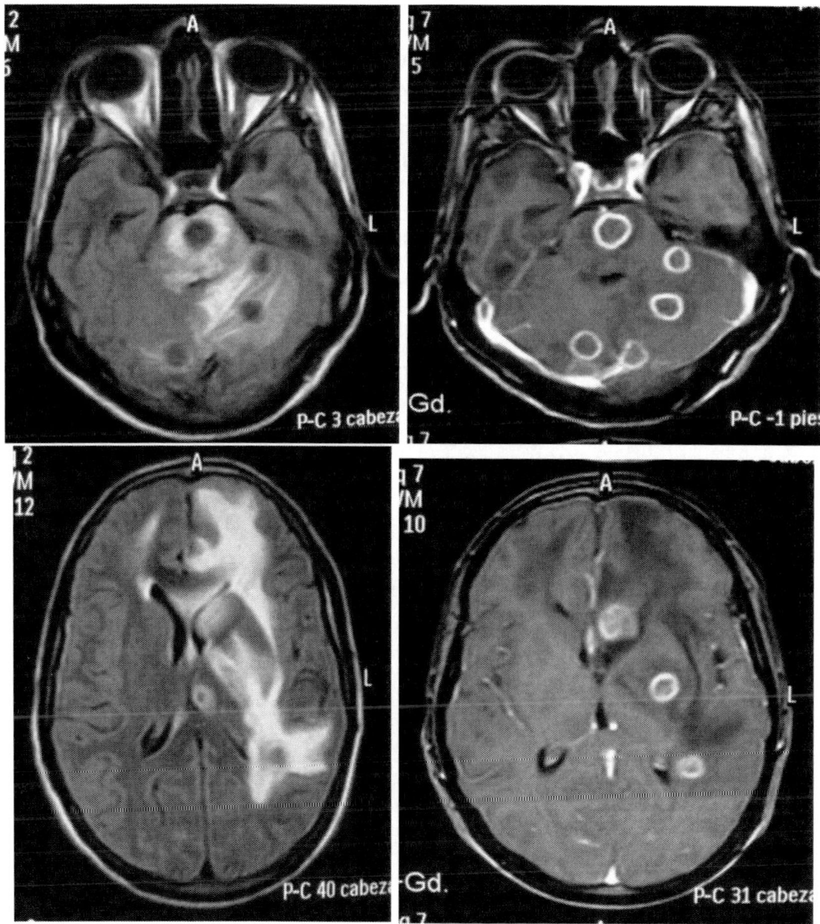

Figure 8A. 25 year old patient, December 2004. Axial T2W and T1W with Gadolinium brain MRI show multiple intraparenchymal granulomas, hypointense in T2 and ring enhancing in T1 images.

A recent meta-analysis concluded than the mean sensitivity and specificity of ADA assays were 79 and 91% respectively. However, publications bias may have resulted in overestimation of diagnostic accuracy [3, 15].

ADA assays may be useful in confirming TBM, but raised levels may also be seen in other CNS disorders. It is not a useful test in HIV-positive patients [15]. The detection of MT DNA in CSF samples using polymerase chain reaction (PCR) is a widely used diagnostic method. A meta-analysis found a sensitivity of 56% for the commercial testing of PCR and a specificity of 98%. Sensitivity of PCR testing is higher in culture-positive patients [9, 32, 33].

A recent study observed that CSF filtrates contain a substantial amount of MT DNA [9], suggesting that the filtrates and not sediments are likely to reliably provide a PCR-based diagnosis [34]. The PCR test does not replace the conventional tests such as microscopic, culture and biopsy. Results of nucleic acid amplification test should be interpreted in conjunction with conventional test and clinical data [35]

When the results of smear, culture and PCR were compared in a study carried out in India, it was found that the sensitivity of CSF microscopy was only 3.3%, culture 26.7%, CT scan 60%, and PCR 66.7% [32-36]. These results, as well as others, show an increase in the percentage of positivity when diagnostic tests are combined, especially if they are conducted in various CSF samples [3, 9]. Greater positivity has been found when various diagnostic tests are associated with a higher degree of neurological deterioration of the patients, especially if, in these patients, the smear results are positive [3]. Similar results have been found with new methods such as Anti-Bacillus Calmette-Guérin antibody-secreting cell detection in CSF by an enzyme-linked immunospot assay, valuable because of its high degree of sensitivity [9, 37, 38]. The WHO for the first time endosed a new rapid diagnostic test for TB, the GenXpert MTB/RIF(Cepheid). The GenXpert is a desktop machine that simultaneously detects the presence of MTB and rifampicine resistance in specimens using nested real-time PCR [15].

Neuroimaging

Computed Tomography (CT) and Magnetic Resonance Imaging (MRI) are valuable in diagnosis, evaluation of complications and prognosis of TBM. All patients with TBM must have imaging studies with and without contrast. Characteristic changes in CTimages include cysternal effacement, leptomeningeal enhancement, (Figure 7), cerebral infarcts (Figures 8 a, b), granulomas (Figure 9) and hydrocephalus (Figures 10 a, b c). In our series [13] of 310 patients, we found hydrocephalus in 35% of patients who had a complete recovery or mild or moderate deficit, in 64.3% of patients who had severe deficit, and in 73.7% of patients who died. Cysternal effacement was found in 19% of the patients who recovered without deficit or mild or moderate deficit, 42.9% that had severe deficit, and in 49.1% of those who died. Leptomeningeal enhancement showed 19.9% of patients who recovered without deficit or had mild deficit, 34.6% who had moderate deficit, 46.4%

who had severe deficit, and 50.9% of patients who died. Infarcts were present in 17.7% of patients without deficit or with slight or moderate deficit, 39.3% of patients who showed severe deficit, and in 42.1% of patients who died, with superficial location in 40% of cases and deep location in 60%. We found granulomas in 18% of cases. Prognosis is worse if the patient has hydrocephalus, cysternal effacement and deep infarcts.

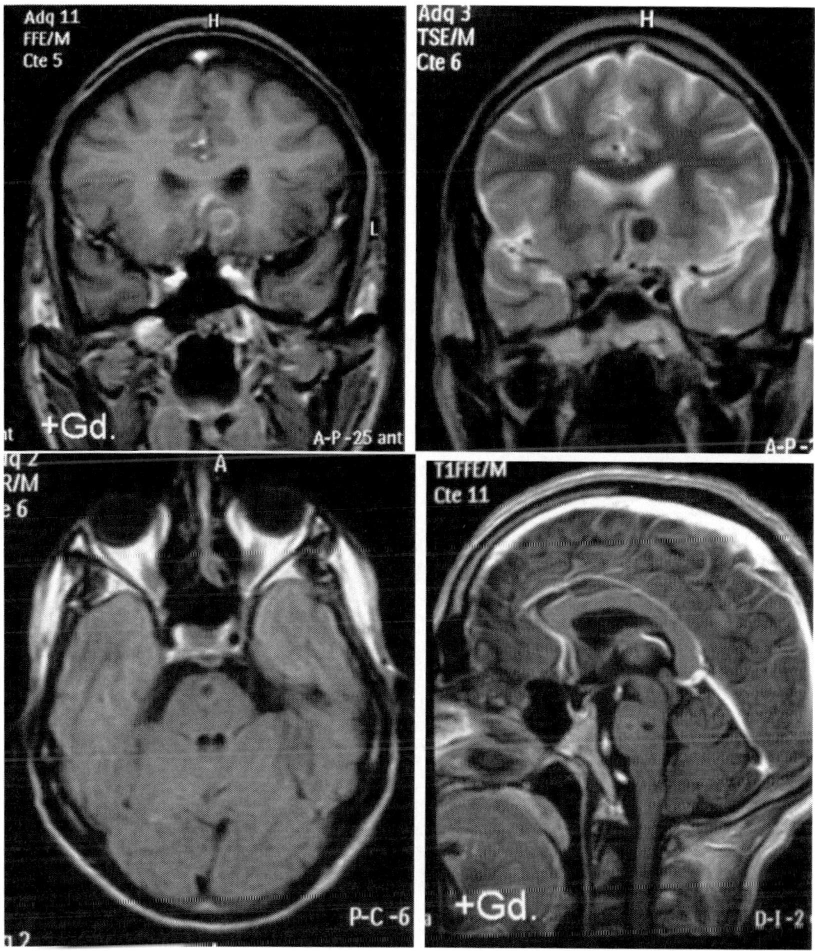

Figure 8B. 25 year old patient, August 2005. Coronal, axial and sagital T1W with Gadolinium, T2W, FLAIR and TW1 with Gadolinium show thalamic and pontine granulomas. The last one near a small infarct secondary to vasculitis.

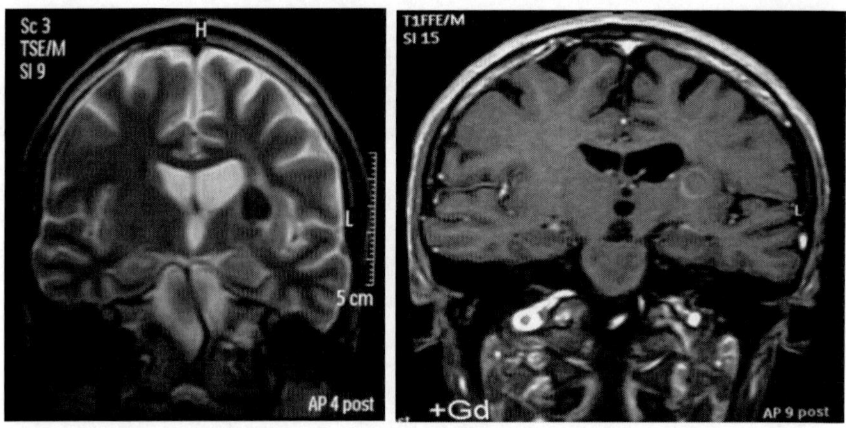

Figure 9. 34 year old patient. Coronal T2W and T1W with Gadolinium brain MRI show a hypointense in T2 and ring enhansing in T1W striatum granuloma.

Therapy

The prognosis for TBM, which was almost invariably fatal in the early twentieth century, has improved substantially after the advent of streptomycin in the forties and isoniazid in the fifties [7]. The drugs for tuberculosis have been divided by general consensus into first-line drugs, which include isoniazid, rifampicin, ethambutol, pyrazinamide, and streptomycin, and second-line drugs which include4-aminosalicylic acid, ethionamide, cycloserine, and some aminoglycosides and quinolones [7]. Fluoroquinolones (levofloxacin, gatifluoxacin, and moxifluoxacin) are antimicrobial agents used in drug-resistant tuberculosis. Substitution of older fluoroquinolones, especially ciprofloxacin, into a regime meant for treating drug-resistant tuberculosis resulted in a higher rate of relapse [9]. The main therapeutic principle in TBM is that the antituberculous therapy must be administered as soon as the disease is suspected. Postponing start-up of treatment even for a few hours or days may increase morbi-mortality [3, 5]. When TBM is suspected, because antituberculous therapy is not particularly toxic in the short term, empirical treatment should be started as early as possible to reduce morbidity and mortality [3, 5, 7]. Antimicrobial therapy often must be administered empirically, much before bacteriological confirmation. Most first-line drugs, except ethambutol, penetrate the CSF satisfactorily. The concentration of isoniazid and pyrazinamide in CSF is below the minimum inhibitory concentration. These two drugs and streptomycin do not penetrate uninflamed meninges. Rifampicin is bactericidal, whereas streptomycin and ethambutol are tuberculostatic [7, 9, 38]. The concentration of rifampicin and streptomycin three hours after administration is over the inhibitory concentration but declined later. Corticosteroids had no effect on CSF penetration of antituberculous drugs [39].

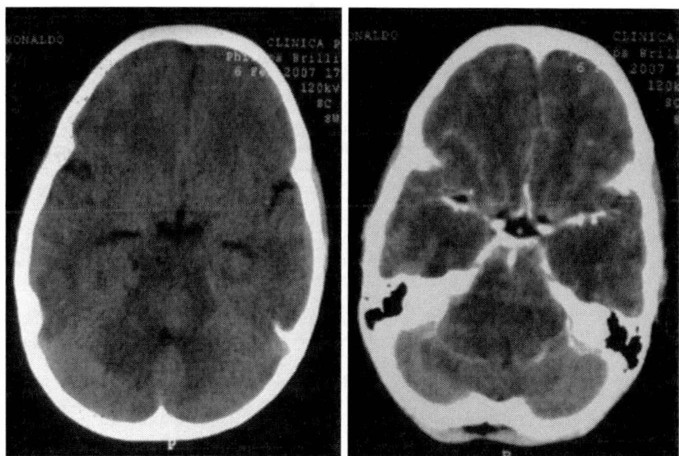

Figure 10A. 8 year old patient, February 2007. CT scan without and with contrast show diffuse brain hypodensity and basal meningeal enhancent with signs of vascular entrapment and vasculitis. Mild hyrocephalus.

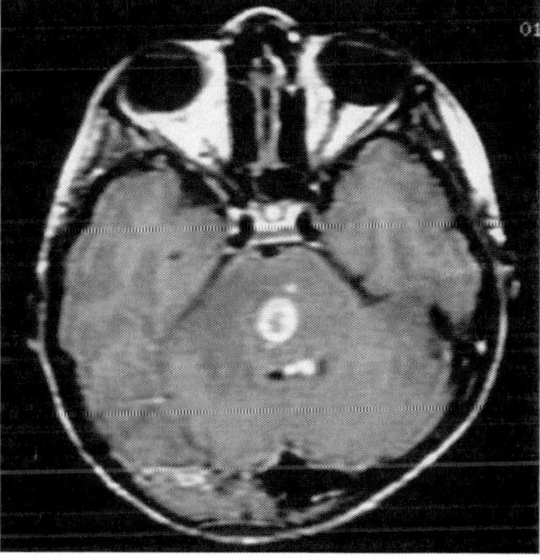

Figure 10B. 8 year old patient, February 2007. Axial T1W brain MRI with Gadolinium, eight months after treatment, shows pontine and intraventricuar granulomas.

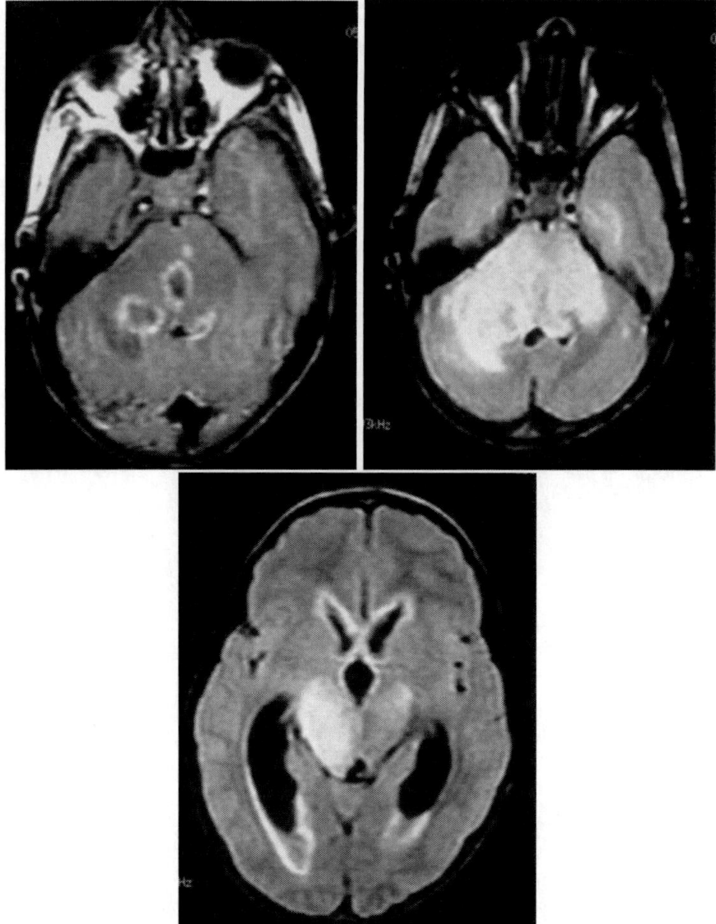

Figure 10C. 8 year old patient, March 2007. Axial T1W with Gadolinium and FLAIR brain MR images show multiple pontine and right middle cerebellar peduncle granulomas with severe edema, ependymitis and hydrocephalus with caseum in the occipital horns of the lateral ventricles.

There is no consensus about the optimal duration of TBM treatment. Widely accepted regimes include short 6 months and long 9 to 18 months duration treatment [3, 5, 7]. A 6-month therapeutic regime has proven to be effective with a morbidity and mortality ratio similar to that found in the longer course therapies [3, 5]. Whenever drug resistance is unlikely, a regime with three drugs, rifampicin, isoniazid, and pyrazinamide (all daily), for two months and two drugs (isoniazid and rifampicin daily) for four additional months (10 months in children) seems a reasonable option in the adult [3, 5,

7]. The low toxicity of these drugs and their powerful sterilizing action has led them to be considered as the most important. Some authors recommend the initial use of a fourth drug for TBM, usually ethambutol [7], especially when there is pulmonary tuberculosis. The World Health Organization (WHO) recommends a category-based treatment for TBM [40]. The antituberculous treatment regime is divided into two phases: an initial intensive phase and a second phase. In the intensive phase, the antituberculous therapy regime includes a combination of four first-line drugs: isoniazid, rifampicin, streptomycin and pyrazinamide. The intensive phase continues for two months [9]. A regime without streptomycin is equally effective [5]. In the second phase, a two-drug regime, with isoniazid and rifampicin, is administered for at least four months [5, 9, 13]. In patients with TBM, the second phase is usually extended to 7 or 12 months, especially for patients with delayed response as judged by mycobacterial cultures that remain positive or by inadequate resolution of symptoms and signs and in those cases where CSF continues with pleocytosis or proteins with amounts that are higher than normal [3, 7, 9, 13]. Some experts have recommended a twice-a-week drug regime [9]. Antituberculous therapy in HIV-positive patients remains the same as for HIV-negative patients [7, 9, 42, 42].

Multidrug-resistant TBM should be considered if there is a history of contact with a patient of multidrug-resistant pulmonary tuberculosis, in patients that have previously received more than one incomplete or irregular treatment for pulmonary tuberculosis or TBM or a poor clinical response to antimicrobial therapy despite adequate treatment [43-45] When the patient that is being treated shows, after six months, pathological CSF, the treatment must be extended until CSF falls within normal limits [3, 4, 13]. The regimes recommended for this type of patient must include at least five drugs. The treatment regime must include drugs that the patient has not received before and to which the patient's organism is susceptible. The regime must also include an injectable medication. In the five-drug regime, one of the antituberculous drugs must be fluoroquinolone [9]. The initial six-month phase must be followed by a continuation phase from 12 to 18 months. With the appearance of tuberculosis that is multidrug-resistant to at least isoniazid and rifampicin, any fluoroquinolone and at least one of the injectable drugs like amikasin, kanamycin and capreomycin [9] must be administered.

Immune reconstitution inflammatory syndrome (IRIS) is a potentially life-threatening condition which is seen in HIV-positive patients of tuberculosis [9]. This complication can be seen within three months after starting highly active antiretroviral therapy. IRIS is characterized by improvement in CD4 cell counts. Its differential diagnosis includes failure of the antituberculous treatment, drug reactions, paradoxical clinical deterioration, antituberculous and antiretroviral drugs interaction with loss of antiretroviral efficacy and alternative opportunistic conditions. IRIS may be severe enough to cause death [9, 39, 46, 47].

The administration of antituberculous and antiretroviral drugs at the same time may produce significant interactions between the drugs, which could lead to a loss of effectiveness of the antiretroviral drugs [9]. In general, treatment with antituberculous drugs is reasonably well tolerated by patients. Hepatitis induced by antituberculous drugs does not reach unacceptable levels [5]. For successful treatment, it is a challenge to handle this complication. If a patient develops hepatitis induced by antituberculous drugs, treatment must be suspended temporarily. After liver function tests show that the patient has recovered, the therapeutic regime must start up again, with the gradual and progressive reintroduction of each one of the drugs until the complete dose is being administered. In TBM, it is advisable that at least two drugs that have no hepatotoxic effects (like ethambutol and streptomycin) be continued [3, 5, 9, 48]. There is need to improve the pharmacokinetic and pharmacodynamic principles and CSF penetration of antituberculous drugs [15].

Therapy with corticosteroids continues to be controversial despite the recent Cochrane review, which recommends to use corticosteroids routinely in TBM, because they substantially reduce death and neurological disability after treatment. Nevertheless, our results do not show any benefits [3, 5, 13]. The Cochrane review [49] refers to a well-designed, randomized, double-blind, placebo-controlled trial conducted in Vietnam. In this study, the patients were randomly assigned to receive either dexamethasone or a placebo. Patients with grade II or grade III of British Medical Research Council (BMRC) TBM received intravenous dexamethasone for four weeks and then oral dexamethasone for four weeks. Patients with grade-I disease received two weeks intravenous dexamethasone and the four weeks of oral therapy. Nine months of follow-up treatment with dexamethasone was associated with a significantly improved survival. However, treatment with corticosteroids did not alter the combined outcome of death and severe disability. Grade-I patients showed a slightly significant benefit for the combined outcome [50]. Furthermore, the patients who received corticosteroids had slightly fewer adverse events as a result of the antituberculous drugs than those in the placebo group [9]. Some authors have shown that dexamethasone affects the prognosis of TBM, possibly because it reduces the incidence of hydrocephalus and cerebral infarct [51]. The mechanism of action of the corticosteroids in

TBM is not known. A recent study showed that dexamethasone decreased CSF matrix metalloproteinases-9 concentrations early in the course of the treatment. The authors suggest that this could be one of the mechanisms whereby the corticosteroids would improve the prognosis in TBM [52]. These results are not similar to those we have found in our series of 310 patients where steroids were used in patients with severe loss of consciousness, bilateral motor impairment, vasculitis, spinal tuberculosis and intracranial tuberculomas with intracranial hypertension and focal deficit [3, 5, 13]. In our patients (Table 1), we do not use corticosteroids on grade-I patients or most grade-II patients. Nevertheless they had a better prognosis in terms of death and neurological disability. Corticosteroids were administered to 80 patients (25.8%). It was found, after completing the antituberculous treatment, that 49% of patients recovered completely, of which 43.4% were in grade I, 34.9% in grade II, and 21.7% in grade III. We found severe sequels in 17.9% of grade-I patients, in 32.1% of grade-II patients, and in 50.0% of grade-III patients. 5.3% of grade-I patients, 38.6% of grade-II patients, and 56.1% of grade-III of our patients died. The combined outcome of death and severe disability in our patients was 9.4 % of the cases that started the treatment in grade I, in 36.4 % of the patients that started in grade II and in 54.1 % of the patients who started treatment in grade III. In our study the use of corticoids was associated with a poor prognosis, may be because we used steroids in patients with severe tuberculosis, who have poor prognosis.

Tuberculomas, Tuberculous Abscesses, and Spinal-cord Involvement

Tuberculous granulomas (tuberculomas) are composed of a central zone of solid caseation necrosis that is surrounded by a capsule of collagenous tissue, epithelioid cells, multinucleated giant cells and mononuclear inflammatory cells. They contain a few bacilli in the necrotic center. Outside the capsule, there is parenchymal edema and astrocytic proliferation [7]. Tuberculomas can be found in the cerebrum, cerebellum, subarachnoid space, subdural space or epidural space [7]. In children, the tuberculomas tend to be infratentorial, whereas in adults they tend to be supratentorial. They can coexist with meningitis in 10 to 15% of cases and be multiple in more than one third of the patients [7, 53, 54].

Clinical course is subacute or chronic, lasting weeks or months. Chest X-rays suggest pulmonary TB in 30 to 80% of patients. Mantoux test is positive in up to 85% of the patients. CSF findings are unremarkable or show slight abnormalities, and bacteriology is usually negative. The diagnosis is therefore made on the basis of neuroimaging findings, PPD results and response to antituberculous therapy [7, 55, 56]. In CT scans, tuberculomas may appear as solid enhancing rim enhancing or mixed lesions. On occasions, there is a central calcification surrounded by a hypodense area with a peripheral rim enhancement (target sign), a highly suggestive pattern, although not pathognomonic for TB [7, 57]. On MRI images, tuberculomas appear as isointense to gray matter on T1-weighted images and many have a slightly hyperintense rim. Noncaseating lesions are bright on T2-weighted images. Caseating tuberculomas vary from isointense to hypointense on T2-weighted images and exhibit rim enhancement [7]. Tuberculomas may show a variable degree of perilesional edema, which is usually more prominent in the early stage [7, 13]. Differential diagnosis includes neoplasias and other granulomatous processes such as sarcoidosis and parasitic diseases such as cysticercosis and toxoplasmosis [7].

With medical treatment, tuberculomas usually diminish in size and are completely resolved in three months. In some cases, the resolution extends over several months and sometimes years, leading to residual calcification. The medical treatment, which is similar to TBM, is initially recommended and surgery advised only if there is intolerable intracranial hypertension or failure of medical treatment. Mortality with chemotherapy regimes is inferior to 10% [7, 9, 13, 58].

The development or growth of brain tuberculomas in the weeks following the start of antimicrobial treatment has been recognized as a paradoxical response [58-61]. This paradoxical reaction appears in 2% of cases with TBM [58]. The paradoxical response may be interpreted as a clinical deterioration occurring weeks after the start of therapy, sometimes associated with an increase in CSF pleocytosis, more commonly lymphocytic. Response may change transiently in the direction of polymorphonuclear predominance. This is interpreted as a hypersensitivity reaction to the massive release of mycobacterial proteins [7, 58, 60]. This response appears from three weeks up to various months after starting the antituberculous therapy [7, 58]. Enlargement of existing cerebral tuberculomas along with an aggravation of anterior cerebral artery vasculitis, despite the appropriate treatment, has been described [62]. The paradoxical reaction also occurs during the antituberculous treatment in HIV-positive patients when antiretroviral therapy restores immune function [7]. Although there are no controlled studies, it has been suggested that therapy with corticosteroids improves prognosis [7, 58].

When the caseous core of a tuberculoma liquefies, a tuberculous abscess results. The abscesses are usually larger and less frequent than tuberculomas. Frequently multiloculated, they often exert a greater impact on mass and edema. In contrast to the solid caseation and few organisms seen in tuberculomas, the abscess is formed by pus, where many bacilli can be found. The appearance recalls a typically pyogenic abscess. A TB abscess has a faster course and its clinical manifestations are more evident than in tuberculomas, with high fever, headache, focal neurological signs and impaired consciousness. The appropriate therapy includes antituberculous chemotherapy and surgical removal when necessary [7, 58, 63].

Spinal cord, roots and vertebral bodies can be affected by tuberculosis. Tuberculous myelitis and radiculomyelitis usually appear as an acute or subacute disease and, in a few cases, chronically. The CSF analysis reveals an increase in protein content with lymphocytic pleocytosis and, in one third of cases, low glucose levels. The mid-thoracic cord is the most commonly affected, followed by the lumbar and cervical regions. MRI shows uptake of the contrast surrounding the spinal cord and the roots and obliterating the subarachnoid space with focal or diffusely increased intramedullary signal on T2 weighted images and variable degrees of edema and mass effect [3, 7]. Spinal meningitis most frequently accompanies intracranial disease, although in certain cases it may occur alone [7]. Rarely tuberculomas occur in the spinal cord, either as intramedullary or dural lesions. Very infrequently intramedullary tuberculous abscesses have been reported. Tuberculous spondilitis affects the thoraco-lumbar region more frequently, with L1 being the most affected. Extended standard antimicrobial therapy is indicated in the spinal lesions. Systemic corticosteroids improve the prognosis of spinal cord lesions. Surgery must be considered on a case-by-case basis.

Prognosis

In TBM, early diagnosis and treatment are important indicators for a better prognosis [3, 5, 13]. Antituberculous treatment prevents death and disability in less than 50% of patients. Impairment of consciousness is significantly associated with mortality and lesser degree of impaired consciousness shows a significantly associated disability [3]. The association between impairment of consciousness and mortality suggest that severe intracranial damage with or without intracranial hypertension or hydrocephalus could be progressive or irreversible. The irreversibility of neurological impairment could be related to ischemia, edema, distortion of the adjacent structures or venous thrombosis. The findings of Ziehl or positive cultures for MT in CSF, accounting for a high concentration of bacilli, has been associated with disease severity and death. In imaging studies, cysternal effacement and exudate in cysterns and sylvian fissures have shown to be associated with disability and mortality [3, 5, 7, 9, 13].

Table 1. Functional prognosis at discharge in 310 patients with TBM

BMRC	Number cases	Total	Mild or moderate	Severe sequelae	Death	Total
	Percent	recovery	sequelae			
BMRC	Count	66	23	5	3	97
Stage I	%	43.2	31.5	17.9	5.8	31.3
BMRC	Count	53	35	9	22	119
Stage II	%	34.9	47.9	32.1	38.6	38.3
BMRC	Count	33	15	14	32	94
Stage III	%	21.7	20.5	50	56.1	30.4
Total	Count	152	73	28	57	310
	%	49	23.5	9	18.4	100

BMRC = British Medical Research Council.
Stages = Stage I, Stage II, Stage III.

Mortality is higher in patients younger than five years of age, older than 50 years of age, and those in whom illness has been present for longer than two months [3, 5, 9]. BMRC staging, which is a scale to assess the severity of TBM, has been related to the prognosis of disease and death. In grade-I patients of the BMRC, disease and death are low, whereas in grade-II patients they are moderate, and in grade-III patients, they are very high (Table 1). Other factors that have been associated with poor outcome include headache, convulsions, motor deficit and cerebral infarcts. High mortality (63.3%) has been reported in HIV patients [9]. Hemiparesis, paraparesis, quadriparesis, aphasia and loss of sight are common neurological impairments among surviving patients. In a study, complete neurological recovery was observed in 21.5% of the surviving patients. However in our series complete recovery was obtained in 43.4%of the patients [13, 64]. The British Medical Research

Council staging has been extensively used to evaluate the disease severity and establish the approximate prognosis of tuberculous meningitis. We compared The BMRC staging with a new scoring. The Score Quito that is easy to apply and is a good predictor of poor outcome [13].

References

[1] Sinner, S. W. (2010). "Approach to the diagnosis and management of tuberculous meningitis."*Curr Infect Dis Rep* 12 (4): 291-298.

[2] Marais, S., G. Thwaites, et al. (2010). "Tuberculous meningitis: a uniform case definition for use in clinical research." *Lancet Infect Dis*10 (11): 803-812.

[3] Alarcón F, Cevallos N, Narvaes (1994). "Meningitis Tuberculosa: Un Viejo y Nuevo problema", *Rev Ecuat Neurol.* 3:39-54.

[4] Moreira, J., F. Alarcon, et al. (2008). "Tuberculous meningitis: does lowering the treatment threshold result in many more treated patients?" *Trop Med Int Health* 13 (1): 68-75.

[5] Alarcon F, Escalante L. (1990). "Tuberculous meningitis short course of chemothery. *Arch Neurol.* 47: 1313-1317.

[6] Thwaites, G. E. and J. F. Schoeman (2009). "Update on tuberculosis of the central nervous system: pathogenesis, diagnosis, and treatment." *Clin Chest Med* 30 (4): 745-754, ix.

[7] Garcia-Monco, J. C. (1999). "Central nervous system tuberculosis."*Neurol Clin* 17 (4): 737-759.

[8] Garg, R. K. and M. K. Sinha (2011). "Tuberculous meningitis in patients infected with human immunodeficiency virus." *J Neurol* 258 (1): 3-13.

[9] Garg, R. K. (2010). "Tuberculous meningitis." *Acta Neurol Scand* 122 (2): 75-90.

[10] Alarcon, F., G. Duenas, et al. (2000). "Movement disorders in 30 patients with tuberculous meningitis."*MovDisord* 15 (3): 561-569.

[11] Alarcon, F., E. Tolosa, et al. (2001). "Focal limb dystonia in a patient with a cerebellar mass."*Arch Neurol* 58 (7): 1125-1127.

[12] Alarcon, F., J. C. Maldonado, et al. (2011). "Movement disorders identified in patients with intracranial tuberculomas."*Neurologia* 26 (6): 343-350.

[13] Alarcon F, Moreira J, Rivera J, Salinas R, Duenas G, Van den Ende J.(2013). Outcome of central nervous system tuberculosis: An alternative scoring system proposal. *Indian J. Tuberc* 60;5-14.

[14] Srikanth, S. G., A. B. Taly, et al. (2007). "Clinicoradiological features of tuberculous meningitis in patients over 50 years of age." *J Neurol Neurosurg Psychiatry* 78 (5): 536-538.

[15] Brancusi F, Farrar J, Heemskerk D.(2012)Tuberculous Meningitis in Adults:a review of a decade of developments focusing on prognostic factors for outcome. *Future Microbiol* 7(9):1101-1116.

[16] Dastur, D. K. and P. M. Udani (1966)."The pathology and pathogenesis of tuberculous encephalopathy."*Acta Neuropathol* 6 (4): 311-326.

[17] Udani, P. M. and D. K. Dastur (1970). "Tuberculous encephalopathy with and without meningitis. Clinical features and pathological correlations."*J Neurol Sci* 10 (6): 541-561.

[18] Alarcon F, Espinosa F, Pesantes B, Banda H, Vinan I. (1988) Encefalopatia tuberculosa: presentacion de un caso y caracteristicas en TAC craneal. *Arch de Neurobiol* (Madrid) 51:213-215.

[19] Molavi, A. and J. L. LeFrock (1985). "Tuberculous meningitis."*Med Clin North Am* 69 (2): 315-331.

[20] Tandon Pn. (1978) Tuberculous meningitis in Vinken PJ, Bruyn, Klawans HL, Eds. *Handbook of Clinical Neurology* Vol 35, Amsterdam: North Holland 1978:195-265.

[21] Kocen RS. (1987) Tuberculosis of the Nervous System. in: Kennedy PGE, Johnson RT, Eds. *Infections of the Nervous System*, London; Butterworth: 23-42.

[22] Hanna, L. S., N. I. Girgis, et al. (1988). "Ocular complications in meningitis: "fifteen years study"." *Metab Pediatr Syst Ophthalmol* 11 (4): 160-162.

[23] Amitava, A. K., S. Alarm, et al. (2001). "Neuro-ophthalmic features in pediatric tubercular meningoencephalitis." *J Pediatr Ophthalmol Strabismus* 38 (4): 229-234.

[24] Misra, U. K., J. Kalita, et al. (2011). "Stroke in tuberculous meningitis." *J Neurol Sci* 303 (1-2): 22-30.

[25] Chan, K. H., R. T. Cheung, et al. (2005). "Cerebral infarcts complicating tuberculous meningitis."*Cerebrovasc Dis* 19 (6): 391-395.

[26] Leiguarda, R., M. Berthier, et al. (1988). "Ischemic infarction in 25 children with tuberculous meningitis."*Stroke* 19 (2): 200-204.

[27] Udani, P. M. (1985). "Management of tuberculous meningitis."*Indian J Pediatr* 52 (415): 171-174.

[28] Kennedy, D. H. and R. J. Fallon (1979). "Tuberculous meningitis."*JAMA* 241 (3): 264-268.

[29] Torok, M. E., T. T. Chau, et al. (2008). "Clinical and microbiological features of HIV-associated tuberculous meningitis in Vietnamese adults."*PLoS One* 3 (3): e1772.

[30] Thwaites, G. E., T. T. Chau, et al. (2004). "Improving the bacteriological diagnosis of tuberculous meningitis."*J Clin Microbiol* 42 (1): 378-379.

[31] Barnes, P. F., A. B. Bloch, et al. (1991). "Tuberculosis in patients with human immunodeficiency virus infection."*N Engl J Med* 324 (23): 1644-1650.

[32] Desai, D., G. Nataraj, et al. (2006). "Utility of the polymerase chain reaction in the diagnosis of tuberculous meningitis."*Res Microbiol* 157 (10): 967-970.

[33] Pai, M., L. L. Flores, et al. (2003). "Diagnostic accuracy of nucleic acid amplification tests for tuberculous meningitis: a systematic review and meta-analysis." *Lancet Infect Dis* 3 (10): 633-643.

[34] Haldar, S., N. Sharma, et al. (2009). "Efficient diagnosis of tuberculous meningitis by detection of Mycobacterium tuberculosis DNA in cerebrospinal fluid filtrates using PCR." *J Med Microbiol* 58 (Pt 5): 616-624.

[35] Desai, M. M. and R. B. Pal (2002). "Polymerase chain reaction for the rapid diagnosis of tuberculous meningitis."*Indian J Med Sci* 56 (11): 546-552.

[36] Thwaites, G. E., M. Caws, et al. (2004). "Comparison of conventional bacteriology with nucleic acid amplification (amplified mycobacterium direct test) for diagnosis of tuberculous meningitis before and after inception of antituberculosis chemotherapy."*J Clin Microbiol* 42 (3): 996-1002.

[37] Thomas, M. M., T. S. Hinks, et al. (2008). "Rapid diagnosis of Mycobacterium tuberculosis meningitis by enumeration of cerebrospinal fluid antigen-specific T-cells."*Int J Tuberc Lung Dis* 12 (6): 651-657.

[38] Moadebi, S., C. K. Harder, et al. (2007). "Fluoroquinolones for the treatment of pulmonary tuberculosis."*Drugs* 67 (14): 2077-2099.

[39] Kaojarern, S., K. Supmonchai, et al. (1991). "Effect of steroids on cerebrospinal fluid penetration of antituberculous drugs in tuberculous meningitis."*Clin Pharmacol Ther* 49 (1): 6-12.

[40] World Health Organization. (2002) *Treatment of tuberculosis: guidelines for national programmes,* 3rd edn. Geneva, Switzerland: World Health Organization.

[41] Berenguer, J., S. Moreno, et al. (1992). "Tuberculous meningitis in patients infected with the human immunodeficiency virus."*N Engl J Med* 326 (10): 668-672.

[42] Garg, R. K. and M. K. Sinha (2011). "Tuberculous meningitis in patients infected with human immunodeficiency virus." *J Neurol* 258 (1): 3-13.

[43] Horn, D. L., D. Hewlett, Jr., et al. (1993). "RISE-resistant tuberculous meningitis in AIDS patient."*Lancet* 341 (8838): 177-178.

[44] Patel, V. B., N. Padayatchi, et al. (2004). "Multidrug-resistant tuberculous meningitis in KwaZulu-Natal, South Africa."*Clin Infect Dis* 38 (6): 851-856.

[45] Byrd, T. F. and L. E. Davis (2007). "Multidrug-resistant tuberculous meningitis."*Curr Neurol Neurosci Rep* 7 (6): 470-475.

[46] Huttner, H. B., R. Kollmar, et al. (2004). "Fatal tuberculous meningitis caused by immune restoration disease." *J Neurol* 251 (12): 1522-1523.

[47] McIlleron, H., G. Meintjes, et al. (2007). "Complications of antiretroviral therapy in patients with tuberculosis: drug interactions, toxicity, and immune reconstitution inflammatory syndrome." *J Infect Dis* 196 Suppl 1: S63-75.

[48] Saukkonen, J. J., D. L. Cohn, et al. (2006). "An official ATS statement: hepatotoxicity of antituberculosis therapy." *Am J Respir Crit Care Med* 174 (8): 935-952.

[49] Prasad K, Singh MB. (2008) *Corticosterois for managing tuberculous meningitis.* Cochrane Database Syst Rev; Art. No.: CD002244.DOI:10.1002/14651858. CD002244.pub 3.

[50] Thwaites, G. E., D. B. Nguyen, et al. (2004). "Dexamethasone for the treatment of tuberculous meningitis in adolescents and adults."*N Engl J Med* 351 (17): 1741-1751.

[51] Thwaites, G. E., J. Macmullen-Price, et al. (2007). "Serial MRI to determine the effect of dexamethasone on the cerebral pathology of tuberculous meningitis: an observational study." *Lancet Neurol*6 (3): 230-236.

[52] Green, J. A., C. T. Tran, et al. (2009). "Dexamethasone, cerebrospinal fluid matrix metalloproteinase concentrations and clinical outcomes in tuberculous meningitis."*PLoS One* 4 (9): e7277.

[53] Arseni, C. (1958). "Two hundred and one cases of intracranial tuberculoma treated surgically." *J Neurol Neurosurg Psychiatry* 21 (4): 308-311.

[54] Jinkins, J. R. (1991). "Computed tomography of intracranial tuberculosis."*Neuroradiology* 33 (2): 126-135.

[55] Loizou, L. A. and M. Anderson (1982). "Intracranial tuberculomas: correlation of computerized tomography with clinico-pathological findings." *Q J Med* 51 (201): 104-114.

[56] Mayers, M. M., D. M. Kaufman, et al. (1978). "Recent cases of intracranial tuberculomas." *Neurology* 28 (3): 256-260.

[57] Whiteman, M. L. (1997). "Neuroimaging of central nervous system tuberculosis in HIV-infected patients."*Neuroimaging Clin N Am* 7 (2): 199-214.

[58] Alarcòn F, Espinosa S, Dueñas G. (2001) Respuesta paradojica y desarrollo de tuberculomas intracraneales durante tratamiento antituberculoso: Reporte de seis casos. *Rev Ecuat Neurol.;* 10: 43-49.

[59] Nicolls, D. J., M. King, et al. (2005). "Intracranial tuberculomas developing while on therapy for pulmonary tuberculosis." *Lancet Infect Dis* 5 (12): 795-801.

[60] Garcia Monco JC, Ferreira E, Gomez-Beldarrain M. (2005) The therapeutic paradox in the diagnosis of tuberculous meningitis. *Neurology*; 65:191-192.

[61] Afghani, B. and J. M. Lieberman (1994). "Paradoxical enlargement or development of intracranial tuberculomas during therapy: case report and review." *Clin Infect Dis* 19 (6): 1092-1099.

[62] Lee, S. I., J. H. Park, et al. (2008). "Paradoxical progression of intracranial tuberculomas and anterior cerebral artery infarction" *Neurology* 71 (1): 68.

[63] Tyson, G., P. Newman, et al. (1978). "Tuberculous brain abscess."*Surg Neurol* 10 (5): 323-325.

[64] Kalita J, Misra UK, Ranjan P. (2007) Predictors of long –term neurological sequelae of tuberculous meningitis: a multivariate analysis. *Eur J Neurol* 2007;14:33-37 (Erratum in: *Eur J Neurol*;14:357.

In: Meningitis
Editor: Anthony L. Shrader

ISBN: 978-1-63117-136-9
© 2014 Nova Science Publishers, Inc.

Chapter 4

Role of Prophylactic Antibiotics in Posttraumatic Meningitis

Carlos Jimenez[1], Luisa Galvis[2] and Manuela Jimenez[3]
[1] Professor, Neurosurgery and Epidemiology,
University of Antioquia, Medellin, Colombia
[2] Resident, Fifth year in Neurosurgery,
University of Antioquia, Medellin, Colombia
[3] Medical Student, Pontificia Bolivariana University,
Medellin, Colombia

Abstract

Traumatic brain injury (TBI) is one of the main causes of admission to neurosurgical services. The occurrence of meningitis after TBI, both closed and open, is rare but its early detection is very critical, since this infection increases morbidity and mortality up to 65%. Pathogens can access the central nervous system by direct contact and invasion through the external boundaries, or rarely from the bloodstream in association with a rupture of the blood-brain barrier. The conditions of barrier disruption are clear in penetrating trauma, and so is the pathogen colonization of the sinuses themselves and contact with the subarachnoid space in closed TBI by means of dural tears. The time elapsed from the moment of the trauma and the diagnosis of meningitis ranges from the first few hours to several years; the period has been reported from 8.4

days to 3.4 years. The mean interval of the different reports is 5-13 days, with 78% of cases diagnosed in the first 15 days. The microbiological agents isolated in post-traumatic meningitis include a wide range of both gram negative and gram positive bacteria, being Streptococcus pneumoniae the agent most commonly reported. The role of antibiotic prophylaxis in patients with CSF leaks has been extensively studied, yet still remains controversial. Published studies suggest that antibiotic prophylaxis does not reduce the risk of meningitis in TBI, both open and closed, since the odds ratio is not statistically significant in most of the clinical trials and prospective studies. The available scientific evidence does not support the use of prophylactic antibiotics to prevent posttraumatic meningitis, even though there are a few reports advocating its utility. When using prophylactic antibiotics, several factors must be considered: in addition to its unproven utility, it should be taken into account that the patient will be exposed to side effects that are not always mild, costs will increase and, last but not least, multiresistant microorganisms will be selected.

General Aspects of Posttraumatic Meningitis

Traumatic brain injury (TBI) is one of the main causes of admission to neurosurgical services. This type of trauma is associated with a host of complications, and among them infection is a serious problem to consider, with the posttraumatic nosocomial meningitis as a potentially devastating event. [1, 2, 3] The occurrence of meningitis after penetrating trauma is rare, with an incidence of 0.38-2.03 %. [1, 2, 3] The diagnosis is sometimes limited because patients involved are frequently critically ill due to comorbidities and existing neurological lesions. However, the importance of the constant search for treatment is that this entity overshadows the long-term prognosis in patients with TBI, increasing morbidity and mortality up to 65%. [1, 4, 5] TBI has many associated conditions that elevate the risk of intracranial infection. [6] The access of infectious agents is related to the mechanism of trauma and the time of onset of signs and symptoms is variable, with many reported cases diagnosed several years after the traumatic event.

Palabiyikoglu et al. proposed several factors associated with posttraumatic meningitis, among them the infection of traumatic wound and the presence of surgical shunt catheters. [7] Some authors found other risk factors, such as cerebrospinal fluid (CSF) leak, male gender, length of surgery and type of intervention. [4] A recent study found that CSF leak is an independent risk factor for posttraumatic meningitis, as well as concomitant infections as sinusitis, otitis and pneumonia. The organisms involved include a wide range of gram-positive and gram-negative bacteria. The flora of the sinuses is mentioned by several authors as the leading source of microorganisms in posttraumatic meningitis. [8] In this manner, there are reports of recurrent meningitis associated to Pneumococcus (30%), isolation of gram-negative bacilli in 50% of patients, gram-positive cocci in 41% and 9% of culture-negative CSF. [1]

When discussing surgery as a risk factor, it has been reported a 5% incidence of postoperative meningitis in TBI, [9, 10] although some authors as Malekpour - Afshar et al. reported an incidence as high as 66.6 % of postoperative meningitis in patients with TBI undergoing surgery. [11] Those who advocate the use of prophylactic antibiotics argue that the infection rate increases up to 10 % if they are not used. [12] Bacterial meningitis as a postoperative complication is one of the most severe conditions. [1, 2, 9, 10, 13] Almost one third of cases are diagnosed in the first postoperative week, but there are reports even years after an intervention. The risk of posttraumatic meningitis in patients undergoing surgery decreases with control of CFS contact with contaminated tissues; here lies the importance of preventing and correcting postoperative CSF leakage. [9, 10, 13] Other factors are infection at the surgical site and procedures longer than four hours. [13]

Mechanisms of Infection

The central nervous system is protected from the entry of microorganisms from the bloodstream by the blood-brain barrier, as well as an external barrier formed by the skull and the leptomeninges. [13] Consequently, pathogens can access the central nervous system by direct contact and invasion through the external boundaries, or through the bloodstream in association with a rupture of the blood-brain barrier. The conditions of barrier disruption are clear in penetrating trauma, and so is the pathogen colonization of the sinuses themselves and contact with the subarachnoid space in closed TBI by means of dural tears. It is well known that infection in penetrating gunshot injury is associated with contamination of brain parenchyma by fragments of skin, scalp, bone and metal and dissemination across the path of the bullet. [14]

Skull fractures, with incidences of 33-89 % in the field of TBI, are frequently accompanied by dural tears and CSF leaks, which then become the entry zone for microorganisms. [1, 6, 9, 10] The CSF leak is diagnosed between 16, 5% - 80 % of cases of TBI [1, 2, 6, 10, 15, 16] Careful medical examination is of great importance because CSF fistulae could be of very low pressure, intermittent or hidden by the presence of hemorrhage. [4, 10, 17] The resolution of CSF leaks can be spontaneous, but the risk of developing meningitis persists as long as the dural tear is present. [1, 18] Other proposed mechanism explaining the CSF colonization of microorganisms involves fractures through sinuses and structures of the middle ear, mastoid cells, even without dural tears. [18, 5, 10] In some cases the connection with spinal fractures and closed pelvic trauma has been described. Pneumocephalus is indicative of dural rupture. Cases involving animal bites or facial burns could be related to retrograde venous flow. [18]

Diagnosis

The diagnostic approach consists of neuroimaging, analysis-culture of CSF and blood cultures. [10, 12] Neuroimaging is indicated in most patients with suspected nosocomial bacterial meningitis, since it leads to identify the presence of foreign bodies, superinfected intracerebral collections associated with trauma, pneumocephalus as well as sinus communication in cases of dural tears. Diagnosis of nosocomial bacterial meningitis is based on the results of the CSF culture, but because of the length of time necessary to obtain cultures and the possibility of negative results by prior antibiotic use, CSF cytochemical analysis regains importance, especially to determine cell counts and biochemical tests for glucose and protein, as well as Gram staining. [13] CSF cell counts can be useful but have low sensitivity and specificity in injured patients, [13, 19] so it is possible to find normal cell counts in 22% of patients undergoing placement of external ventricular catheters, with a similar number of patients presenting with pleocytosis but negative cultures. [19] Among patients with suspected postoperative meningitis, the entity known as aseptic meningitis, the result of an inflammatory response to breakdown products in the blood, can occur in up to 70% of cases. [20] There are additional tests for patients at high risk of developing meningitis, e.g., in whom a surgical procedure has been performed or who have suffered moderate to severe TBI. In patients undergoing neurosurgery, a lactate of 4 mmol per liter or more in the CSF demonstrated a sensitivity of 88%, specificity of 98%, positive predictive value of 96 % and a negative predictive value of 94 % for diagnosis of bacterial meningitis. [12] However, a retrospective review of cases diagnosed with meningitis showed an error of up to 50 % for predicting neuroinfection with this cut-off of lactate concentration. [12, 13]

Other tests are being evaluated, such as C-reactive protein in serum or CSF and measurement of serum procalcitonin, even though these are unspecific markers of inflammatory response associated with bacterial infection processes and its utility in the diagnosis of posttraumatic meningitis requires additional tests. [10, 12, 13]

Signs and Symptoms

Signs and symptoms of posttraumatic meningitis are variable; most patients are critically ill and have evidence of CNS compromise. [1, 3, 4, 18, 9, 21] Fever has been reported in 86 -100% of cases, [3, 4, 18, 21] deterioration in the state of consciousness in 97-100% of cases. [3, 18] The change in mental status can be rapidly progressive. [5] The triad of fever, neck stiffness and changes in the state of consciousness is common and its presentation in a patient evolving from a TBI should always lead to the suspect of meningitis. [10] Headache is also reported in 57-86% of cases; however, several authors emphasize the ambiguity and the lack of specificity of this symptom, due to the affected mental state in head injured patients. [18, 21, 10]

Course

The time elapsed from the moment of the trauma and the diagnosis of meningitis ranges from the first few hours to several years; the period has been reported from 8.4 days to 3.4 years. [18, 6, 9, 17, 5, 16, 22] The mean interval of the different reports is 5 13 days, [1, 6, 9, 7, 17] with 78% of cases diagnosed in the first 15 days. [23] Most of posttraumatic infections secondary to gunshot wounds appear early: 55% in the first 3 weeks and 90% by the sixth week. [15, 24, 14] Rarely, infections have been reported up to 15 years after the trauma. [17]

Microbiological Diagnosis

The microbiological agents isolated in post-traumatic meningitis include a wide range of both gram negative and gram positive bacteria. Streptococcus pneumoniae is the agent most commonly reported in most of the reports, isolated in 52-100% of cases. [4–7, 9, 15, 18, 21, 23] The second group in frequency consists in Staphylococcus aureus, Streptococcus and other species of cocci; [9, 18, 21] Gram negative are reported with a frequency ranging from 17-100%, especially in cases of penetrating injury or prolonged hospitalization [1–3, 9, 17] (1, 2, 3, 8, 11). The isolated germs include Escherichia coli, [1, 2, 4, 9] Klebsiella pneumoniae [1, 2, 9] Neisseria meningitides, [18] Haemophilus influenzae, [18, 21] and Pseudomonas aeruginosa. [2, 21, 23]

The diagnosis of posttraumatic meningitis is confirmed with positive CSF cultures, which are present in 73-100% of patients. [1, 4, 6, 9, 18] Culture-negative CSF is reported in 27-41% of cases. Additionally, positive blood cultures were present in up to 86% of patients, much more in infections related to Pneumococcus. [6, 21]

Treatment

During the pre-antibiotic era most patients with gunshot penetrating injury developed intracranial infection, and studies of that time reported infection in 58.8% of cases; [14] according to some authors this high incidence has declined, following the use of prophylactic antibiotics. [25] Despite this concept, Jimenez et al. recently reported that the use of prophylactic antibiotics was not associated with decreased incidence of infection in patients with gunshot penetrating TBI. [14]

Empirical selection of antibiotics should be guided by the clinical presentation, the prevalence of microorganisms at the specific healthcare center and the drug penetration through the blood-brain barrier. In patients with posttraumatic meningitis and closed TBI empirical antibiotics against pneumococcus are recommended; [9] however, the growing pneumococcal resistance to antibiotics indicates the need for alternative antibiotic therapy. [20] In patients with penetrating TBI, prolonged hospitalization, use of prophylactic antibiotics or late onset of meningitis, definitive therapy should include broad-spectrum antibiotics, considering the high risk of infection with multiresistant or gram negative organisms. [4, 9, 17, 20]

Antibiotic therapy for patients with either penetrating TBI or closed injury accompanied by basilar skull fractures, and diagnosed during prolonged postoperative hospitalization, should consist of vancomycin in combination with cefepime, ceftazidime or meropenem. [26] Individual susceptibility profiles reported at the health care facility must be considered when choosing the antibiotic agent. Meropenem is the empirical agent of choice in these group of patients, because of its safety profile, particularly respect to the occurrence of seizures. [13, 26] Empirical treatment for patients with diagnosis of skull base fractures should consist of vancomycin plus a third generation cephalosporin (either cefotaxime or ceftriaxone). [1, 13] In the therapeutic arsenal for empirical treatment of posttraumatic meningitis linezolid and daptomycin are good alternatives, with proven efficacy in many cases of staphylococcal meningitis. [13, 26] Furthermore, the linezolid has a unique feature of favorability regarding penetration (approximately 80% at equilibrium) in neurosurgical patients. [25]

Respect to immunization, there are not reported benefits of routine application of a pneumococcal vaccine to prevent meningitis in patients with TBI. [23]

Beyond the use of antibiotic therapy, there are other measures for the treatment of posttraumatic meningitis, especially if recurrent, and surgical correction of a CSF leak is one of the most important one, especially if there is a suspected pneumocephalus associated with basilar fractures. [20] There are different surgical approaches, taking into account the possibility of repair via rhinoseptal approach; however, it is clear that the correction of the sphenoid defect must be one of the objectives when it is present, and thus the neurosurgical exploration for dural defects regains importance because of the amount of exposure and the possibility to correct those defects. [14] Similarly, the endoscopic approach is a therapeutic option when the bone defect is related to a CSF leak, approachable by this method. [12, 14, 16] The timing for surgical intervention in patients with persistent posttraumatic CSF leak is not clear. The literature recommends considering surgical management with traumatic fistulae that persist more than 24 hours. [22] On average, patients are taken to surgery after 40-45 days of the trauma for correction of a CSF leak; however, in patients with high debit clinically evident fistulae the timing for surgery has a mean interval of 16 days. Patients with recurrent meningitis and hidden fistulae are taken later to the operating room, with reported cases of surgery as late as six years after trauma. [22]

Outcome

The long-term prognosis is variable. The reported mortality is diverse, with frequencies ranging from 0 to 65%. [1, 2, 4, 6, 7, 9, 10, 17, 18, 21] The most common complications include pneumonia, deafness, anosmia, major neurological deficits. The series aimed at determining the functional prognosis, showed that 47-75% of survivors remained independent at discharge. [1, 3] neurological deficits are associated with damage to the brain parenchyma, because many patients have worsening prognosis (death or a low score on functional outcome scales) attributable to the presence of post traumatic meningitis. [23]

Role of Antibiotic Prophylaxis in Preventing Posttraumatic Meningitis

General Aspects

It is generally accepted that between 7% and 30% of all patients with posttraumatic CSF leakage will develop meningitis, and this rate increases as long as the CSF fistula persist. [5, 10, 11, 14, 24] The role of antibiotic prophylaxis in patients with CSF leaks has been extensively studied, yet still remains controversial. However, as patients with persistent fistulae are at increased risk of meningitis, the efficacy of antibiotic prophylaxis may be even higher in this subgroup of patients. [1, 9, 10, 27] One of the topics discussed is the need for a large volume of patients to show a benefit of routine use of prophylactic antibiotics. [27] Patients included in most studies had received prophylaxis in traumatic CSF leaks, even if the fistula disappeared in the first 24 hours, which is quite questionable, because this is precisely the group of patients who seem to benefit less from using prophylactic antibiotics. Some authors recommend antibiotic prophylaxis for a period of between 3 to 14 days , even a week after resolving CSF leaks. [28] In a meta-analysis Villalobos et al. found several retrospective studies with different antibiotics schemes, including first-generation penicillin, cephalosporin, chloramphenicol, amides, sulfa, in combination with gentamicin and penicillin. In more recent studies, third-generation cephalosporins were used. In two prospective studies ceftriaxone in combination with ampicillin / sulfa and penicillin were used. In all studies, antibiotic treatment was started within the first 72 hours of hospitalization and continued for three days to one week after CSF leakage was resolved. [28] An analysis of 1241 patients from 12 published studies suggest that antibiotic prophylaxis does not reduce the risk of meningitis, since the OR (1.15) was not statistically significant. [28]

Acute bacterial meningitis is defined as inflammation of the leptomeninges (pia-arachnoid) caused by bacteria; it is an important cause of death and disability, especially in developing countries, and its case-fatality rate remains quite high, between 10 and 30%. [29] On the other hand, there are two types of traumatic brain injury (TBI) who have a high risk of meningitis; these are open or penetrating and closed injury with accompanying basilar skull fractures. [30] Bacterial meningitis is one of the most serious complications associated with TBI with basilar skull fractures; its incidence is between 0.2 and 17.8%, and may reach 50% when there is a cerebrospinal fluid (CSF) leak. [3, 4] Several studies have found that basilar skull fractures are present in 7 to 15.8% of patients with closed TBI, who in turn present a CSF leak in 2 to 20.8% of cases; [30] this CSF leak is indicative of rupture of the dura and connects the subarachnoid space with nasopharyngeal flora, as well as of sinuses and middle ear, which is ultimately the mechanism by which bacterial meningitis takes place. Clinical signs that may lead the physician to suspect a basilar skull fracture in a patient who is evolving from a closed TBI are: CSF rhinorrhea or otorrhea, bilateral periorbital ecchymosis, Battle sign, peripheral facial paralysis or tympanic rupture, with blood inside the external auditory canal, deafness or hearing loss, vestibular dysfunction and evidence of anosmia. The diagnosis should be confirmed by performing a high resolution brain computarized tomography (CT) scan. [30]

Regarding risk factors for posttraumatic meningitis, it is clear that the presence of a penetrating injury, such as those caused by firearms, [14] or the identification of basilar skull fracture with CSF leak, [31] increase significantly its incidence. Others propose that the mere presence of a pneumocephalus is an independent risk factor for infection, considering this as an indirect indicator of a non-visible CSF leak, and the patient should be treated as having one. [31] On the other hand, the diagnostic approach of a patient suspicious of having posttraumatic meningitis is complex, because bacteriological demonstration of infection is not possible all the times; authors have no unified diagnostic criteria and some have recommended the classification in two types: verified (culture-positive) and probable but not verified (culture-negative but CSF analysis compatible with bacterial infection). [32] The mechanism by which the pneumocephalus occurs after the TBI is the presence of a dural tear, especially in fractures that involve the sinuses, mastoid air cells and petrous pyramid of the temporal bone. In relation to gunshot penetrating trauma, the rate of meningeal and brain tissue infection seems to be quite high, with reports of incidence as high as 58.8% in the early twentieth century before the first World War, to that reported by Jimenez et al. in 2013, with an incidence of 25% of meningitis and other intracranial infections in a cohort study that included 160 patients. [14] These authors found that the risk factors for meningeal or brain tissue infection were: persistent bone or metal fragments in brain tissue after surgical procedure, projectile trajectory through the paranasal sinuses or oral cavity, and prolonged hospital stay. [14]

Prophylactic Antibiotics

The use of prophylactic antibiotics in the field of neurosurgery has been highly controversial. There is a growing concern about the emergence of multiresistant pathogens due to their widespread use. [30] Some authors refuse the indication and use of prophylactic antibiotics, not only in head injury but also in other contexts; for example, they have not demonstrated their efficacy in preventing the highly contagious community acquired meningococcal meningitis, since the incidence in all the contacts is similar regardless of whether they are given or not an antibiotic regimen against *Meningococcus*. [29] Something similar occurs in case of prevention of infection in patients undergoing craniotomy, in whom the germs most frequently involved are *Staphylococcus* and *Enterobacteria*. [33]

The debate persists in relation to the utility of prophylactic antibiotics in posttraumatic meningitis, with authors who favor their use, considering that the presence of open fractures or CSF leaks are associated with high risk of meningeal infection, while others propose that all that prophylactic antibiotics do is select the most pathogenic antibiotic resistant germs, making the infection more difficult to treat. [34] This concept has been derived from the analysis of multiple studies, some of them focused on the use of antibiotics prior to craniotomy. [34] The review of the different studies, including at least two metaanalysis, leads to the conclusion that there is no agreement on this issue, and also makes clear the poor statistical power of the published reports. [31, 35]

The first studies that mentioned the utility of prophylactic antibiotics to prevent meningitis in patients with TBI followed the experience of Cairns during the World War II, and were published in 1947; after a relative initial enthusiasm, gradually its use almost disappeared in the following decades, since many researchers advocated methodological weakness in previous reports, and some authors even referred worsening of patients due to the side effects of antibiotics, without being observed any decrease in the incidence of meningitis. [34] One issue in analyzing the different reports is the publication bias, since studies showing favorable results with prophylactic antibiotics tend to be more frequently published. [34] The other problem with the evaluation of the effectiveness of prophylactic antibiotics lies in the definition of meningitis, because many researchers consider the diagnosis of meningitis even in cases of culture-negative CSF, following the concept proposed by Finlyson and Penfield in 1941 and known as "aseptic meningitis", with CSF changes that, on the contrary, many authors consider induced by the bleeding secondary to trauma instead of an infectious origin. [34]

The most frequently used prophylactic antibiotics for meningitis in patients with TBI are ampicillin, cephalothin, penicillin and ceftriaxone; all of them have been administered intravenously for a time period of between seven and ten days. [30] In a review of the Cochrane Collaboration published in 2010, by Ratilal-Costa-Sampaio,[30] which included five clinical trials and 17 descriptive and case series studies, there were no differences in the incidence of meningitis among patients with and without the use of prophylactic antibiotics (Peto OR 0.68, 95% CI 0.28 to 1.65). In the same way, when the final outcome between the two groups (prophylactic antibiotics vs. no prophylactic antibiotics) was all-cause mortality, there were no statistically significant differences (Peto OR 1.76, 95% CI 0.41 to 7.60); finally there were no differences when mortality associated with meningitis was evaluated (Peto OR 1.03; 95% CI 0.06 to 16.44). [30] At the end the authors conclude that prophylactic antibiotics showed no effect in preventing meningitis in patients with closed TBI in whom a basilar skull fracture was demonstrated, with or without associated CSF leak. Additionally, the authors recommend that at the moment of using prophylactic antibiotics in these patients, it has to be taken into account factors such as the risk of adverse effects and the costs associated with its use. [30] Finally, they draw the attention about the necessity for randomized controlled clinical trials (RCT), with better methodological quality than the previously published reports. In a subsequent review by the same authors, also conducted for the Cochrane Collaboration and published in 2011, in which was evaluated the utility of prophylactic antibiotics in the prevention of posttraumatic meningitis in patients with closed TBI and basilar skull fractures, [36] the authors focused only on prospective clinical trials available at the time of the review, and they included a total of five studies with 208 patients. They found that only one study (Klastersky, 1976) was randomized, controlled and double-blind. The remaining four were not placebo-controlled and the outcomes were not blindly measured. In this review some biases were found in relation to treatment allocation, blinding and incomplete evaluation of the results [36]. Anyway, there were not statistically significant differences in the incidence of meningitis among the patients with and without prophylactic antibiotics (Peto OR 0.69, 95% CI 0.29 to 1.61); they found no significant differences for all-cause mortality (Peto OR 1.68, 95% CI 0.41 to 6.95) or for meningitis-related mortality. (Peto OR 1.03; 95% CI 0.14 to 7.40) [10] Once again the authors conclude that the scientific evidence up to the date of publication of the work in 2011 does not support the use of prophylactic antibiotics in preventing meningitis in patients with TBI and associated basilar skull fractures, with or without CSF leak. [36]

In a meta-analysis published in 1998 by Villalobos et al., [28] in which the authors included 12 studies with 1.241 patients, it was found that the odds ratio (OR) for the prevention of meningitis using prophylactic antibiotics in patients with closed TBI accompanied by basilar skull fractures was not statistically significant, since it had a value of 1.15 but the confidence interval included 1.0. Same thing happened when the group of patients with CSF leak was analyzed separately. In other study conducted in Iran and published in 2004, [4] Eftekhar et al. found that there were no statistically significant differences in relation to the administration of ceftriaxone at a dose of 1 g IV every 12 hours for five days to a group of 56 patients with closed TBI and pneumocephalus compared to no use of antibiotics in 56 patients with the same diagnosis. However, should be taken into account that this was not a blinded study and also it had no placebo-control group; in addition, the diagnosis of meningitis was not dependent on the isolation of a germ in the CSF culture.

A Cochrane review published in 2007, (1) which evaluated the effectiveness of prophylactic antibiotics for posttraumatic meningitis when patients were in high risk of infection, defining this risk in terms of the presence of basilar skull fractures accompanied by CSF leak, revealed that to date four controlled clinical trials were published, with a total of 207 patients included; the reviewed studies had as a common denominator the poor methodological quality, and, when they were subjected to a metaanalysis, there was no statistically significant difference in the frequency of meningitis among the groups with and without prophylactic antibiotic therapy (Peto odds ratio [OR] 0.68, 95% CI 0.28-1.65 or), since mortality (Peto OR 1.76, 95% CI 0.41-7.60) or non-CNS infection (Peto OR 0.62, 95% CI 0.16-2.41). [29] Until

2007, systematic reviews and meta-analyzes indicated that there is insufficient evidence to support or refute the use of prophylactic antibiotics in the approach of meningitis in patients with TBI and basilar skull fractures; there is the necessity of a major outcome studies of excellence in its methodology. [29]. When it has to do with open TBI secondary to gunshot wounds, in a cohort type prospective study, developed by Jimenez et al. and published in 2013, it was found that the use of prophylactic antibiotics did not reduce the incidence of meningeal or brain tissue infection [14].

Conclusion

The available scientific evidence does not support the use of prophylactic antibiotics to prevent posttraumatic meningitis, even though there are a few reports advocating its utility.

To date, the published clinical trials suffer from insufficient sample size, which gives them low statistical power, and most show considerable biases, such as poorly defined diagnostic criteria, inadequate blinding when the results are evaluated, null randomization and some of them even lack of a control group. When choosing the use of prophylactic antibiotics, several factors must be considered: in addition to its unproven utility, it should be taken into account that the patient will be exposed to side effects that are not always mild, costs will increase and, last but not least, multiresistant microorganisms will be selected.

References

[1]   S. P. Baltas I, Tsoulfa S, "Posttraumatic meningitis: Bacteriology, hydrocephalus, and outcome," *Neurosurgery*, vol. 35, no. (3), September, pp. 422–427, 1994.

[2]   F. J. Buckwold, R. Hand, and R. R. Hansebout, "Hospital-acquired bacterial meningitis in neurosurgical patients.," *J. Neurosurg.*, vol. 46, no. 4, pp. 494–500, 1977.

[3]   B. J. Taha JM, Haddad FS, "Intracranial infection after missile injuries to the brain: Report of 30 cases from the Lebanese conflict," *Neurosurgery*, vol. 29, no. December (6), pp. 864–868, 1991.

[4]   K. A. Lau YL, "Post-traumatic meningitis in children," *Injury*, vol. 17, pp. 407–409, 1986.

[5]   S. M. Michowiz SD, Wald U, "Brucella melitensis meningitis following head trauma," *Infection*, vol. 15, pp. 130–131, 1987.

[6]   S. J. Hand WL, "Posttraumatic bacterial meningitis.," *Ann Intern Med*, vol. 72, pp. 869–874, 1970.

[7]   S. Hoşoğlu, C. Ayaz, a Ceviz, B. Cümen, M. F. Geyik, and O. F. Kökoğlu, "Recurrent bacterial meningitis: a 6-year experience in adult patients.," *J. Infect.*, vol. 35, no. 1, pp. 55–62, Jul. 1997.

[8]   et al. Gehanno P, Loundon N, Barry B, "Meningitis associated with ear and nose diseases. Me´d Mal Infect.," *Med. Mal. Infect.*, vol. 26, pp. 1049–1052, 1996.

[9]   S. J. Jones SR, Luby JP, "Bacterial meningitis complicating cranial-spinal trauma.," *J. Trauma*, vol. 13, pp. 895–900, 1973.

[10] W. H. Thomas M, "Atypical meningitis complicating a penetrating head injury.," *J Neurol Neurosurg Psychiatry*, vol. 54, pp. 92–93, 1991.

[11] R. Malekpour-afshar, S. Karamoozian, and H. Shafei, "Post Traumatic Meningitis in Neurosurgery Department Kerman University of Medical Sciences , Kerman , Iran Department of Neurosurgery , Kerman Medical School , Kerman University of Medical Sciences , Afzal Research Center , Kerman , Iran Afzalipour Hospit," *Am. J. Infect. Dis.*, vol. 5, no. 1, pp. 21–25, 2009.

[12] D. H. Van, E.B., F.P. Bakker, "Infections after craniotomy: A retrospective study.," *J. Infect.*, vol. 12, pp. 105–109, 1986.

[13] N. Nesseler, Y. Launey, and P. Seguin, "Nosocomial bacterial meningitis.," *N. Engl. J. Med.*, vol. 362, no. 14, pp. 1346–7; author reply 1347–8, Apr. 2010.

[14] C. M. Jimenez, J. Polo, and J. A. España, "Risk factors for intracranial infection secondary to penetrating craniocerebral gunshot wounds in civilian practice.," *World Neurosurg.*, vol. 79, no. 5–6, pp. 749–55, 2013.

[15] E. M. Frazee RC, Mucha P, Farnell MB, "Meningitis after basilar skull fracture: Does antibiotic prophylaxis help?," *Postgr. Med*, vol. 83, pp. 267–274, 1988.

[16] J. M. Stillwell M, Hoge C, Hoyt N, "Posttraumatic meningococcal meningitis: Case report.," *J. Trauma*, vol. 31, pp. 1693–1695, 1991.

[17] W. W. Crawford C, Kennedy N, "Cerebrospinal fluid rhinorrhoea and Haemophilus influenzae meningitis 37 years after a head injury," *J Infect 1994;2893-97.*, vol. 28, pp. 93–97, 1994.

[18] E. Appelbaum, "Meningitis following trauma to the head and face.," *JAMA*, vol. 173, pp. 116–120, 1960.

[19] R. P. Schade, J. Schinkel, F. W. C. Roelandse, R. B. Geskus, L. G. Visser, J. M. C. van Dijk, M. C. Van Dijk, J. H. C. Voormolen, H. Van Pelt, and E. J. Kuijper, "Lack of value of routine analysis of cerebrospinal fluid for prediction and diagnosis of external drainage-related bacterial meningitis.," *J. Neurosurg.*, vol. 104, no. 1, pp. 101–8, Jan. 2006.

[20] G. Rocchi, E. Caroli, E. Belli, M. Salvati, M. Cimatti, and R. Delfini, "Severe craniofacial fractures with frontobasal involvement and cerebrospinal fluid fistula: indications for surgical repair.," *Surg. Neurol.*, vol. 63, no. 6, pp. 559–63; discussion 563–4, Jun. 2005.

[21] B. J. Wilson NW, Copeland B, "Posttraumatic meningitis in adolescents and children," *Pediatr. Neurosurg.*, vol. 91, no. 16, pp. 17–20, 1990.

[22] Q. L. Friedman JA, Ebersold MJ, "Post-traumatic cerebrospinal fluid leakage.," *World J. Surg.*, vol. 25, pp. 1062–1066, 2001.

[23] D. N. D. Oka, A. V. Tokpa, and A. B. Jibia, "Posttraumatic Meningitis: Four Case Reports and a Review of the Literature," *Neurosurg. q*, vol. 23, no. 1, pp. 44–48, 2013.

[24] R. C. Kingsland, "Actinobacillus ureae meningitis: Case report and review of the literature.," *J Emerg Med*, vol. 13, no. 5, pp. 623–627, 1995.

[25] S. Van Der Geest, P. W. De Leeuw, and E. E. Stobberingh, "Indications for antibiotic use in ICU patients: a one-year prospective surveillance, Journal of Antimicrobial Chemotherapy," *J. Antimicrob. Chemother.*, vol. 39, pp. 527–535, 1997.

[26] A. R. Tunkel, B. J. Hartman, S. L. Kaplan, B. A. Kaufman, K. L. Roos, W. M. Scheld, and R. J. Whitley, "Practice Guidelines for the Management of Bacterial Meningitis," *Clin. Infect. Dis. an Off. Publ. Infect. Dis. Soc. Am. Infect Dis*, vol. 19129, no. August, pp. 1267–1284, 2004.

[27] B. Ratilal, J. Costa, and C. Sampaio, "Antibiotic Prophylaxis for Preventing Meningitis in Patients With Basilar Skull Fractures: A Systematic Review," *J Evid-based Med*, vol. 6, no. 8, 2006.

[28] T. Villalobos, C. Arango, P. Kubilis, and M. Rathore, "Antibiotic prophylaxis after basilar skull fractures: a meta-analysis.," *Clin. Infect. Dis.*, vol. 27, no. 2, pp. 364–9, Aug. 1998.

[29] K. Prasad and N. Karlupia, "Prevention of bacterial meningitis: an overview of Cochrane systematic reviews.," *Respir. Med.*, vol. 101, no. 10, pp. 2037–43, Oct. 2007.

[30] R. Bo, J. Costa, and C. Sampaio, "Antibiotic prophylaxis for preventing meningitis in patients with basilar skull fractures (Review)," no. 1, 2010.

[31] B. Eftekhar, M. Ghodsi, F. Nejat, E. Ketabchi, and B. Esmaeeli, "Prophylactic administration of ceftriaxone for the prevention of meningitis after traumatic pneumocephalus: results of a clinical trial.," *J. Neurosurg.*, vol. 101, no. 5, pp. 757–61, Nov. 2004.

[32] C. N. Meyer, I. S. Samuelsson, M. Galle, and J. M. Bangsborg, "Adult bacterial meningitis: aetiology, penicillin susceptibility, risk factors, prognostic factors and guidelines for empirical antibiotic treatment.," *Clin. Microbiol. Infect.*, vol. 10, no. 8, pp. 709–17, Aug. 2004.

[33] D. De Bels, a-M. Korinek, R. Bismuth, D. Trystram, P. Coriat, and L. Puybasset, "Empirical treatment of adult postsurgical nosocomial meningitis.," *Acta Neurochir. (Wien).*, vol. 144, no. 10, pp. 989–95, Oct. 2002.

[34] F. G. Barker, "Efficacy of prophylactic antibiotics against meningitis after craniotomy: a meta-analysis.," *Neurosurgery*, vol. 60, no. 5, pp. 887–94; discussion 887–94, May 2007.

[35] B. Eftekhar, M. Ghodsi, A. Hadadi, M. Taghipoor, S. Z. Sigarchi, V. Rahimi-Movaghar, E. S. Kazemzadeh, B. Esmaeeli, F. Nejat, A. Yalda, and E. Ketabchi, "Prophylactic antibiotic for prevention of posttraumatic meningitis after traumatic pneumocephalus: design and rationale of a placebo-controlled randomized multicenter trial [ISRCTN71132784].," *Trials*, vol. 7, p. 2, Jan. 2006.

[36] R. Bo, J. Costa, C. Sampaio, and L. Pappamikail, "Antibiotic prophylaxis for preventing meningitis in patients with basilar skull fractures (Review)," no. 8, 2011.

In: Meningitis
Editor: Anthony L. Shrader

ISBN: 978-1-63117-136-9
© 2014 Nova Science Publishers, Inc.

Meningitis Associated with Autoimmune Diseases

Masakazu Nakamura[1] and Masaaki Niino[2,]

[1]Department of Neurology, Hokkaido University Graduate School of
Medicine, Sapporo, Japan
[2]Department of Clinical Research, Hokkaido Medical Center,
Sapporo, Japan

Abstract

Autoimmune diseases can be organ-specific or systemic, with the
latter type affecting many different organs. As a neurological
complication particularly of systemic autoimmune diseases such as
systemic lupus erythematosus and Behçet disease, meningitis can occur at
any stage of the disease and can be a main symptom as well as a
prognostic factor. Mechanisms of meningitis associated with autoimmune
diseases are still unclear, although disease-specific autoantibodies and
cytokines are considered to play critical roles in the pathogenesis.

In meningitis associated with autoimmune disease, the main
symptoms are fever, headache, nausea, and neck stiffness, which are
similar to those presenting in infectious, malignant, or other forms of
meningitis. The clinical course is also varied and can be acute, chronic, or

*Tel.: +81 11 6118111, Fax: +81 11 611 5820, E-mail: niino@hok-mc.hosp.go.jp

recurrent, and thus is not that helpful in differential diagnosis. It is extremely difficult to prove that comorbid autoimmune disease is a direct cause of meningitis, but the exclusion of other etiologies is essential for accurate diagnosis. Such diagnosis is made from a full set of physical and laboratory data and imaging findings. Furthermore, in patients with autoimmune diseases treated with immunosuppressive or immunomodulating agents, it is often difficult to differentiate meningitis associated with autoimmune disease from infectious or drug-induced meningitis.

Treatment options for meningitis related to autoimmune diseases are targeted to suppress activated autoimmunity, and immunosuppressive agents such as pulsed cyclophosphamide and steroids are often recommended for acute relapse or severe forms of the disease. Recently, new agents such as monoclonal antibodies have been tested in cases involving the central nervous system; however, their efficacy remains unclear.

In this chapter, we discuss autoimmune diseases with which meningitis can occur, and review the clinical features, diagnosis, and treatments for meningitis associated with autoimmune diseases.

Introduction

Meningitis can be observed during the course of autoimmune diseases. Diagnosis of meningitis is based on the signs of meningeal irritation, inflammatory findings in cerebrospinal fluid (CSF), and meningeal enhancement on brain magnetic resonance imaging (MRI), and this diagnostic procedure is the same as for meningitis of other etiologies. However, the typical symptoms of meningitis are often absent in autoimmune diseases, making meningitis associated with autoimmune diseases often difficult to diagnose. In some autoimmune diseases, meningitis is a critical complication of the central nervous system (CNS). In patients treated with corticosteroids and/or other immunosuppressants, the differential diagnosis of co-morbid meningitis in autoimmune diseases or infections is important for treatment, but it can be extremely difficult to diagnose. Carcinomatous meningitis can occur in patients treated long term with corticosteroids and other immunosuppressants. Moreover, nonsteroidal anti-inflammatory drugs (NSAIDs), which are often used for patients with autoimmune diseases, may also induce meningitis.

In patients with autoimmune diseases, meningitis can occur due to various causes, such as autoimmunity, infection, medication, or malignancy, with different prognoses. Here, we discuss autoimmune diseases in which meningitis can occur and review the clinical features, diagnosis, and treatment of meningitis associated with these diseases.

BehÇet Disease

Behcet disease (BD) is an idiopathic vascular inflammatory multisystem disease characterized by recurrent oral and genital ulcers and uveitis [1, 2]. There is a geographical variation in its prevalence, with a higher rate in the region along the Silk Route from the Mediterranean basin to Japan and a very low rate in Europe and North America. There is a similar geographical variation in HLA-B51 sensitivity, which is strongly associated with this disease [3].

BD manifests neurological complications mainly in the CNS, which is termed neuro-Behçet disease (NBD) [4, 5]. Neurological complications commonly develop a few years after the onset of BD [6-8], although they can occur as the first symptom of BD [6, 9, 10]. Such complications are clinically important as they are the most serious causes of long-term mortality and morbidity in BD [11], even though the prevalence of NBD is low (around 5% in general, and about 10% in the area of highest prevalence of BD along the Silk Route [6-8, 12, 13]). CNS involvement shows two forms. The first is parenchymal type, which is characterized by any objective CNS abnormalities detected on neurological examination, neuroimaging, or CSF tests. These findings are based on small-vessel disease. Cases of parenchymal type BD mainly show brainstem or pyramidal tract syndrome, and many cases present neurological symptoms as acute attacks, while some present with a chronic progressive course [6, 7, 14]. The second is non-parenchymal (vascular) type, characterized by extra-axial large vessel disease (cerebral venous sinus thrombosis, intracranial aneurysm, or extracranial aneurysm/dissection). These cases have limited symptoms such as headache from intracranial hypertension [4, 5, 15].

Meningitis is usually observed as a comorbid finding with 70–80% of parenchymal type cases [6, 7, 14], whereas it occurs less frequently as the sole presentation (only 4 of 50 NBS cases in a UK report [7] and 1 of 200 cases in a Turkish study [6]). Fever and headache are common symptoms in the exacerbation of parenchymal neurological involvement. However, these symptoms are often observed as part of the comorbid systemic features, which include malaise, orogenital ulcers, skin lesions, and uveitis [7, 10, 12]. Headache due to direct neurological involvement is present in around 10% of patients, whereas primary headache (e.g., migraine and tension-type headache) accounts for 70% of headaches [16-20]. In addition, non-structural headache associated with systemic exacerbations of BD is frequently reported, and there is no evidence of direct CNS involvement in this type of headache [16, 18, 19]. In the CSF, white blood cell counts are significantly raised with a neutrophil-dominant pattern in the early stage, changing into a lymphocyte-dominant pattern in the later stage of meningitis [5], and the protein level is also moderately increased, often above 1 g/dl [6-8, 14]. Raised white blood cell counts and protein levels in the CSF are thought to be poor prognostic factors [6, 7, 14]. Elevated interleukin-6 (IL-6) levels in the CSF are considered to be a feature of NBD, and are associated with disease activity and prognosis [21-24]. In addition, it has been suggested that CSF IL-6 levels higher than 20 pg/ml indicate a poor long-term prognosis [21]. There are no specific treatments for meningitis in NBD, and treatments are the same as for other forms of NBD (Table 1).

Parenchymal acute attacks in NBD are usually treated with corticosteroids, either high-dose intravenous methylprednisolone or oral prednisolone [4, 5]. These treatments are followed by a slow tapering of oral prednisolone over 2 to 3 months in order to avoid early relapse [25]. The efficacy of other immunosuppressive or immunomodulatory drugs, such as colchicine, azathioprine, cyclophosphamide, methotrexate, chlorambucil, mycophenolate mofetil, tacrolimus, and thalidomide, has been demonstrated for some systemic manifestations, but not fully for neurological involvement of BD [4, 5].

It has been reported that chlorambcil improves CSF pleocytosis with meningoencephalitis in BD [26], although its efficacy for NBD is still unclear. Cyclosporin A, commonly used for ocular involvement of BD, is not recommended for NBD due to neurotoxicity and the risk of inducing neurological complications [27-31]. For NBD with a chronic progressive course, weekly oral methotrexate was shown to slow the rate of deterioration in one study [23], but this finding has not been confirmed. Recently, the efficacy of monoclonal anti-tumor necrosis factor (TNF) -α antibody for NBD has been reported [32-37]. An expert panel recommended that the anti-TNFα antibody infliximab could be used for patients refractory to therapies with pulse cyclophosphamide and prednisolone, or those who relapse while on maintenance therapy with azathioprine or prednisolone [38].

Table 1. Treatments for CNS manifestation of autoimmune diseases focusing on aseptic meningitis

autoimmune disease	meningitis related to autoimmune disease	1st-line therapy for acute attacks and maintenance	2nd-line therapy for refractory cases or to avoid adverse effects of corticosteroids	3rd-line therapy for severe or more refractory cases	Expected therapy (efficacy needs to be confirmed)
Behçet disease (BD)	neuro-Behçet disease (NBD) (parenchymal type involving aseptic meningitis)	corticosteroids	switch to or add immunosuppressants (cyclophosphamide, azathioprine, etc.) except for cyclosporin A	monoclonal anti-TNFα antibody (infliximab)	
systemic lupus erythematosus (SLE)	neuropsychiatric SLE (NPSLE) involving aseptic meningitis	corticosteroids	combination with immunosuppressants (intravenous cyclophosphamide is recommended)	high-dose cyclophosphamide with autologous hematopoietic stem cell transplantation	mycophenolate mofetil monoclonal anti-CD20 antibody (rituximab)
rheumatoid arthritis (RA)	rheumatoid pachymeningitis (RPM)	corticosteroids	combination with immunosuppressants (cyclophosphamide or methotrexate)		
	rheumatoid leptomeningitis (RLM)	corticosteroids	switch to or add in immunosuppressants (cyclophosphamide, cyclosporin A, etc)	monoclonal anti-TNFα antibody (infliximab, etc)	
Sjögren s syndrome (SjS)	neurologic manifestation involving aseptic meningitis	corticosteroids	switch to or add in cyclophosphamide (intravenous administration is common)	other immunosuppressants (combination with corticosteroids or monotherapy)	monoclonal anti-CD20 antibody (rituximab or epratuzumab) monoclonal anti-TNFα antibody (infliximab or etanercept)
sarcoidosis	neurosarcoidosis (NS) involving aseptic meningitis	corticosteroids	combination with immunosuppressants	TNFα antagonist (pentoxifylline or thalidomide) monoclonal anti-TNFα antibody (infliximab)	

Most NBD patients have been shown to recover well from acute attacks, whereas after more than 10 years from onset of neurologic involvement, 20–50% of patients have at least mild neurological disabilities due to repeat relapses or conversion to a progressive disease course [4, 6-8, 14]. Optimal treatments for parenchymal neurological complications including meningitis are expected to be established to prevent neurological sequelae.

Systemic Lupus Erythematosus

Systemic lupus erythematosus (SLE) is a chronic autoimmune disease characterized by multi-organ involvement, including the peripheral and central nervous systems. These neurological syndromes in SLE patients in which other causes have been excluded are termed neuropsychiatric syndromes of SLE (NPSLE), and classified by the American College of Rheumatism into 19 categories, one of which is aseptic meningitis [39]. NPSLE develops before or around the time of diagnosis of SLE [40]. NPSLE has been reported in 23.3–95.0% of patients with SLE, whereas the rate of aseptic meningitis is consistently very low (0–2%) [41-46]. These neurological manifestations are associated with a poor prognosis [40, 47, 48].

The pathogenesis of NPSLE is considered to be due to many factors, including microangiopathy, premature atherosclerosis, autoantibody production, and intrathecal production of pro-inflammatory cytokines [49, 50]. Accumulating evidence indicates that collapse of the integrity of the blood–brain barrier, mainly due to the up-regulated expression of adhesion molecules on vascular endothelial cells, which is attributed to pro-inflammatory cytokines and autoantibodies, is critical for development of NPSLE [51-53]. Anti-phospholipid antibodies were reported to be associated with thrombotic events, including stroke [54] and cognitive dysfunction [55-57] in SLE patients. Antibodies against ribosome P were reported to be associated with psychosis, depression, and NPSLE activities [58-60]. There have been some reports that anti-glutamate receptor antibodies might be associated with cognitive dysfunction and/or depression, but this is still controversial [61, 62].

Meningitis associated with SLE generally consists of aseptic meningitis as NPSLE (see above), infectious septic or aseptic meningitis, or drug-induced aseptic meningitis (DIAM; especially by NSAIDs). Many defects of the immune system in SLE, such as complement protein and receptor deficiency, impaired spleen function [63-65], and decreased T cell count and function [66,

67], increase the risk of infection. In addition, corticosteroids and immunosupressants commonly used for SLE lead to immune insufficiency [66-68]. Thus, infectious meningitis should always be kept in mind in the diagnosis of SLE patients. The infectious agents *Listeria monocytogenes* [69], *Mycobacterium tuberculosis* [70], and *Cryptococcus neoformans* [71] should be considered. It has been reported that SLE patients are predisposed to NSAID-related meningitis (DIAM) [72]. In particular, they tend to show some specificity for ibuprofen, and a specific cell–cell-mediated immunity to ibuprofen has been reported [73, 74]. DIAM is also important in the differential diagnosis of meningitis in SLE in those patients treated with NSAIDs.

Signs of meningeal irritation (fever, headache, nausea, vomiting, and neck stiffness) usually accompany these three types of meningitis, and the nature of these symptoms does not differ among the three types [75, 76]. Thus, it is important to have an understanding of chronic headaches without meningitis (such as migraine, tension-type headache, or intracranial hypertension) which are often observed in NPSLE (20–70%) [41-44, 46].

In the CSF of SLE cases with aseptic meningitis, mild pleocytosis predominantly of lymphocytes is usually observed, and mildly to moderately elevated protein is reported [45, 77-79]. Although there have been some reports of decreased CSF glucose levels in SLE, whether these findings are disease-specific has not yet been confirmed [77, 80]. Higher cell counts, higher protein level, and lower glucose level might suggest septic meningitis; however, these findings are not always crucial differential points [75, 76]. Culture, stain (gram, india ink, and acid-fast bacillus), latex agglutination test (for some bacteria), and PCR (containing virus) are essential to diagnose infectious meningitis. It is difficult to discriminate with CSF findings between NPSLE and DIAM following administration of NSAIDs as the cause of aseptic meningitis.

The optimal treatment for aseptic meningitis with SLE as NPSLE is not clear. Antibiotics and a stress dose of steroids are often provided as primary treatment while waiting for laboratory tests to confirm or rule out septic meningitis [75, 76]. It seems that full recovery could be expected in aseptic meningitis as NPSLE, whereas septic meningitis is likely to cause psychosis (cognitive dysfunction and depression) and some focal signs [75, 76]. For DIAM, it takes a few days to recover after drug withdrawal, and this course is helpful to diagnosis [72]. Steroids alone or in combination with immunosuppressive therapy are generally recommended for NPSLE [81, 82] (Table 1). Cyclophosphamide is often selected as monthly intravenous (500–

1000 mg/m^2) doses for a 6-month induction period followed by quarterly maintenance doses for a period of 2 years for severe NPSLE [83]. Recently, it has been reported that high-dose cyclophosphamide (200 mg/m^2), with or without autologous hematopoietic stem cell transplantation, is effective for severe NPSLE with other organ involvement [84-88]. Mycophenolate mofetil [89] and rituximab [90, 91] have also demonstrated potentially beneficial effects for SLE.

Rheumatoid Arthritis

Rheumatoid arthritis (RA) is a systemic inflammatory disease in which the joints are the main targets but other organs such as skin, eyes, lungs, blood vessels, and the nervous system may be involved [92]. Pachymeningitis and leptomeningitis are rare CNS complications of RA. These two forms have been discriminated by pathological findings mainly at autopsy, although brain MRI findings or gadolinium enhancement of the dura mater or leptomeninx have recently been used in clinical practice. With regard to symptoms, cranial nerve palsy is characteristic of pachymeningitis, but rarely of leptomeningitis. Pathophysiologic mechanisms and optimal treatments have not yet been fully clarified for either type of meningitis.

i)   Rheumatoid pachymeningitis (RPM)

RPM usually occurs at either the very early or late stage of RA, although it can precede systemic involvement of RA in rare cases. The gender ratio of RPM is equal. Headache and cranial nerve palsy are the main symptoms, and optic nerve involvement is common [93, 94]. Cranial nerve palsy may be caused either by constriction of the nerves by a thickened dura mater or by direct invasion of the inflammatory cells [95]. RPM can also involve hydrocephalus [96], sinus thrombosis [96], transient ischemic attack [97], and intracranial hypotension [98]. Serum rheumatoid factor (RF) is positive in almost all cases [93, 94]. CSF sometimes shows slight pleocytosis, and protein levels are usually elevated [93, 96, 99, 100]. RF in the CSF is often used as a diagnostic marker [99, 101]. The pathology of the biopsied dura mater is most significant for diagnosis. In autopsy specimens, typical rheumatoid nodules are observed as central necrosis surrounded by elongated histiocytes and mononuclear inflammatory cells, whereas biopsy specimens usually show inflammatory cells and fibrosis, and rarely demonstrate rheumatoid nodules probably because of limited tissue sampling [93, 94, 102]. However, MRI is currently the most useful diagnostic tool for RPM. MRI can detect thickening of the dura mater with linear and/or nodular enhancement in idiopathic hypertrophic pachymeningitis [103, 104]. In images without contrast enhancement, the thickened dura is hyperintense on T1-weigted images and hypointense on T2-weighted images [105]. Lesions are likely to be observed in the tentorium involving the falx cerebri and in the tentorium cerebelli [93, 99, 100, 106]. Treatments have not yet been established for RPM. Several decades ago, patients died soon after the onset of RPM in spite of corticosteroid administration [102], which is one of the reasons why many patients were diagnosed by autopsy, although there were some reports of successful treatment by corticosteroids alone [94, 99]. Recently, a combination of corticosteroids and other immunosuppressants (cyclophosphamide [100] or methotrexate [94]) have been indicated as effective treatment options (Table 1). Various prognoses have been reported: almost complete remission and no relapse, some sequalae, and death [93, 94, 102].

ii) Rheumatoid leptomeningitis

Rheumatoid leptomeningitis (RLM) can occur with or without arthritis activity and even as a first symptom of RA. It is noted that meningeal signs are often lacking, but focal signs (e.g., hemiparesis, sensory disturbance of extremities, and seizures) and mental state abnormalities (e.g., disturbed consciousness and delirium) can be present. Serum RF is commonly positive. In the CSF, protein levels are usually elevated, whereas pleocytosis is not usually observed [94, 97, 107-114]. IL-6 levels in the CSF might indicate leptomeningitis [113]. The level of TNF-α in the CSF is controversial; it may [110] or may not be elevated [113] in leptomeningitis. Brain MRI scans reveal leptomeningeal enhancement and high intensity of the subarachnoid spaces on fluid-attenuated inversion recovery (FLAIR) images and diffusion weighted images (DWIs). Leptomeningeal lesions are usually focal, and focal signs are often observed [94, 97, 107-114]. Recently, it was reported that DWIs showed patchy high intensity in pia mater lesions which demonstrated high intensity on FLAIR images [112]. Corticosteroids are the drugs most commonly used to treat leptomeningitis and are effective for many patients. The other immunosuppressants, cyclophosphamide and cyclosporine A, are also reported to be useful [94, 97, 107-114]. Taken together, it seems appropriate to select corticosteroids as a first step in treating rheumatoid meningitis, and change or add other immunosuppressants when corticosteroids are ineffective or have to be stopped because of adverse effects. Anti-TNFα antibodies are expected to cure RLM, whereas pachymeningitis and/or leptomeningitis can occur during the administration of infliximab [107] and adalizumab [115] (Table 1).

Sjögren's Syndrome

Sjögren's syndrome (SjS) is an autoimmune disease characterized by mononuclear infiltration and destruction of the salivary glands leading to xerostomia and xerophthalmia with extraglandular manifestations. SjS is classified as either primary or secondary disease; primary SjS occurs by itself and secondary SjS occurs after the onset of another connective tissue disease.

In SjS, the nervous system can be widely involved with predominance of peripheral nervous system involvement (e.g., polyneuropathy) [116, 117]. The CNS may be involved in up to 20% of cases with SjS [118, 119], with symptoms including myelopathy, multiple sclerosis-like syndrome, aseptic meningitis, or meningoencephalitis [120-122]. Aseptic meningitis and meningoencephalitis have been reported to occur in 20% of patients with CNS complications of SjS [120]. Aseptic meningitis and meningoencephalitis are generally described as acute forms of the disease [120]; subacute or chronic forms have also been reported [123, 124]. In the acute forms, patients show acute-onset signs of meningeal irritation (e.g., fever, nausea, headache, and neck stiffness), sometimes with focal signs and seizures [120]. By contrast, the subacute or chronic forms may lack any or show very few of these meningeal signs, and unexpected symptoms such as cranial nerve palsies and cognitive dysfunction are sometimes presented [123, 124].

Serum tests often reveal anti-nuclear antibodies, RF, anti-Ro/SS-A antibodies, and anti-La/SS-B antibodies. Anti-Ro/SS-A antibodies are considered to be associated with CNS vasculitis in SjS, which might reflect the underlying pathology of meningitis and meningoencephalitis in SjS [119, 121, 125]. The CSF commonly shows mild and moderate pleocytosis, consisting predominantly of polymorphonuclear leukocytes at onset, and then mononuclear cells. Elevated protein levels are also commonly observed. Intrathecal IgG synthesis is also present, and oligoclonal bands are often detected [120, 126]. IL-6 in the CSF might be an indicator of CNS complications of SjS [127]. Specific autoantibodies against recurrent aseptic meningitis in the serum and CSF, recognizing 56-kD and 44-kD nuclear antigens of the CNS neurons, meningeal cells, and cells of other systemic organs, have been reported recently [128]. Brain computed tomography (CT) and MRI may show diffuse leptomeningeal and subarachnoid enhancement, which are not disease-specific findings [120, 123].

Most cases of aseptic meningitis become less severe or completely resolve in response to corticosteroid administration [120, 123, 128] (Table 1). Cyclophosphamide may be recommended as second-line treatment when meningoencephalitis is refractory to corticosteroids. Other immunosuppressants such as azathioprine, cyclosporine, methotrexate, chlorambucil, and tacrolimus may also be effective as monotherapy or in combination with corticosteroids [127, 129-132]. Recently, rituximab has been reported to successfully treat CNS manifestation of SjS [133]. However, B-cell targeted monoclonal antibodies (rituximab and epratuzumab) have not been considered as established treatment options, in contrast to anti-TNFα antibodies (infliximab and *etanercept*) [134]. Plasmapheresis and intravenous immunoglobulin might be worth considering in refractory cases for acute relief of symptoms [134, 135].

Sarcoidosis

Sarcoidosis is a multisystem inflammatory granulomatous disease with the characteristic pathological finding of noncaseating granuloma. The enhanced helper T-cell response is considered to be an important inducer of the disease, although its etiology remains unclear. It usually develops between the ages of 20 and 40 years, and predominantly affects the lungs, anterior uvea, lymph nodes, and skin [136, 137]. The prevalence of sarcoidosis remarkably differs among races; for example, 3–10 per 100,000 population in Caucasians and 35–80 per 100,000 in African Americans in North America [138].

Sarcoidosis can involve both peripheral and central nervous system, and is known as neurosarcoidosis (NS). NS occurs in less than 10% of clinically diagnosed sarcoidosis [139-141]. However, post-mortem studies have revealed that only 50% of cases with NS are diagnosed antemortem, suggesting that subclinical neurological manifestations may be more common [142, 143]. The diagnostic criteria of NS require nervous system pathological findings, and include the Kveim-Siltzbach test (with potential infection transfer and other disadvantages), chest X-ray, and serum angiotensin-converting enzyme (ACE), which are poorly sensitive for NS for defining probable or possible disease [144]. This might be one of the reasons for the low rate of antemortem diagnosis of NS. Some investigators have proposed changing these three criteria to high-definition chest CT, bronchoalveolar lavage with a CD4:CD8 ratio >3.5, and a CD4:CD8 ratio >5 in the CSF. However, more data are needed to confirm the reliability of these criteria [145].

NS manifestations are various, such as cranial neuropathy, myelopathy, seizures, and headaches, and meningitis is also a common complication. Meningeal signs are of major importance in NS; however, meningitis may not be noticed until CSF analysis is conducted, because CSF pleocytosis is observed in as many as 50–70% of NS patients [144, 146-149]. Meningeal enhancement on brain MRI might also be diagnostic evidence in the absence of signs of meningeal irritation. Meningitis as NS may be acute, subacute, or chronic [139-141], and its symptoms may occur in the early course of the disease along with other neurological complications of NS [141, 144], which makes it difficult to diagnose. In the CSF, mild to moderate pleocytosis (10–200 cells/mm^3) with lymphocyte predominance and elevated protein levels (about 2.6 g/l) are commonly observed. Glucose concentration in the CSF is often decreased. Intrathecal IgG production is reported in some patients. Oligoclonal bands are detected in 20–80% of NS patients [144, 146-149]. The CSF ACE level is used as an indicator of diagnosis and disease activity, as ACE is thought to be derived from sarcoid granulomas in CSF [150, 151]. However, the rate of CSF ACE activity in NS is wide and CSF ACE activity could also be observed in other conditions such as infection and malignancy [152].

Taken together, it is currently suggested that CSF ACE activity should be considered as a supplementary index in diagnosis and treatment decision-making. Intracranial NS generally involves the leptomeninges and brain parenchyma attached to the meninges, especially in the basal cisterns. Brain MRI usually shows a nodular or diffuse thickening of the affected meninges, and focal parenchymal lesions in white matter and hypothalamic regions [149, 153]. The latter lesions are hyperintense on T2-weighted images and isointense on T1-weighted images, resembling multiple sclerosis in which the lesions have a mainly periventricular distribution [154]. The concomitant meningeal thickening could be one of the differential diagnostic points [147, 155].

Specific treatments for NS have not been identified, but general treatments are available for sarcoidosis (Table 1). First-line therapy for NS is corticosteroid administration; however, adverse effects with high dosage over a long duration and relapses accompanied by tapering are often experienced [144, 153]. In order to avoid these problems, use of other immunosuppressants should be considered.

Although there have been some reports of NS treated with methotrexate [156, 157], cyclosporine [158], azathioprine [147], mycophenolate mofetil [159], or cyclophosphamide [160], it is unclear which drugs are most efficacious currently. TNFα may have a pathogenic role in NS via formation of granulomas in sarcoidosis, and the TNFα antagonists pentoxifylline and thalidomide are expected to ameliorate the symptoms of NS [153, 161, 162]. A monoclonal antibody against TNFα, infliximab, has shown efficacy for refractory sarcoidosis involving NS [163-168].

Other Autoimmune Diseases

Mixed connective tissue disease (MCTD) [169], Vogt-Koyanagi-Harada disease (KVH) [170], ANCA-associated vasculitis (AAV) [171], primary angiitis of the CNS [172], multiple sclerosis [173], and neuromyelitis optica [174] are also known to exacerbate aseptic meningitis. In each disease, the clinical significance of detecting meningitis is different. In KVH, it is an important finding for diagnosis of the disease. Although meningitis is a rare complication of MCTD and AAV, the complication of meningitis might require more aggressive treatment in those diseases. Wegener's granulomatosis and microscopic polyangiitis in AAV are well known to include hypertrophic pachymeningitis. These two diseases are important in the differential diagnosis of "idiopathic" hypertrophic pachymeningitis [175-178].

Conclusion

Meningitis associated with autoimmune diseases is often difficult to diagnose because autoimmune diseases have various frequencies of occurrence, various clinical courses, and different examination findings as well as a lack of definite evidence to differentiate them from other etiologies such as infection and drugs.

Disease-specific pathogenetic mechanisms of meningitis remain unclear, which has resulted in a lack of consensus regarding the optimal treatment and prognosis. Future studies are required to elucidate the pathogenesis of meningitis associated with autoimmune diseases in order to advance future treatment and improve the prognosis of the patients with autoimmune diseases.

References

[1] Sakane T, Takeno M, Suzuki N, Inaba G. Behcet's disease. *N. Engl. J. Med.* 1999 Oct 21;341(17):1284-91.

[2] Verity DH, Wallace GR, Vaughan RW, Stanford MR. Behcet's disease: from Hippocrates to the third millennium. *Br. J. Ophthalmol.* 2003 Sep;87(9):1175-83.

[3] Verity DH, Marr JE, Ohno S, Wallace GR, Stanford MR. Behcet's disease, the Silk Road and HLA-B51: historical and geographical perspectives. *Tissue Antigens.* 1999 Sep;54(3):213-20.

[4] Siva A, Saip S. The spectrum of nervous system involvement in Behcet's syndrome and its differential diagnosis. *J. Neurol.* 2009 Apr;256(4):513-29.

[5] Al-Araji A, Kidd DP. Neuro-Behcet's disease: epidemiology, clinical characteristics, and management. *Lancet Neurol.* 2009 Feb;8(2):192-204.

[6] Akman-Demir G, Serdaroglu P, Tasci B. Clinical patterns of neurological involvement in Behcet's disease: evaluation of 200 patients. The Neuro-Behcet Study Group. *Brain.* 1999 Nov;122 (Pt 11):2171-82.

[7] Kidd D, Steuer A, Denman AM, Rudge P. Neurological complications in Behcet's syndrome. *Brain.* 1999 Nov;122 (Pt 11):2183-94.

[8] Siva A, Kantarci OH, Saip S, Altintas A, Hamuryudan V, Islak C, et al. Behcet's disease: diagnostic and prognostic aspects of neurological involvement. *J. Neurol.* 2001 Feb;248(2):95-103.

[9] Wechsler B, Dell'lsola B, Vidailhet M, Dormont D, Piette JC, Bletry O, et al. MRI in 31 patients with Behcet's disease and neurological involvement: prospective study with clinical correlation. *J. Neurol. Neurosurg. Psychiatry.* 1993 Jul;56(7):793-8.

[10] Joseph FG, Scolding NJ. Neuro-Behcet's disease in Caucasians: a study of 22 patients. *Eur. J. Neurol.* 2007 Feb;14(2):174-80.

[11] Kural-Seyahi E, Fresko I, Seyahi N, Ozyazgan Y, Mat C, Hamuryudan V, et al. The long-term mortality and morbidity of Behcet syndrome: a 2-decade outcome survey of 387 patients followed at a dedicated center. *Medicine* (Baltimore). 2003 Jan;82(1):60-76.

[12] Al-Araji A, Sharquie K, Al-Rawi Z. Prevalence and patterns of neurological involvement in Behcet's disease: a prospective study from Iraq. *J. Neurol. Neurosurg. Psychiatry.* 2003 May;74(5):608-13.

[13] Serdaroglu P, Yazici H, Ozdemir C, Yurdakul S, Bahar S, Aktin E. Neurologic involvement in Behcet's syndrome. A prospective study. *Arch. Neurol.* 1989 Mar;46(3):265-9.

[14] Al-Fahad SA, Al-Araji AH. Neuro-Behcet's disease in Iraq: a study of 40 patients. *J. Neurol. Sci.* 1999 Nov 30;170(2):105-11.

[15] Siva A, Altintas A, Saip S. Behcet's syndrome and the nervous system. *Curr. Opin. Neurol.* 2004 Jun;17(3):347-57.

[16] Borhani Haghighi A, Aflaki E, Ketabchi L. The prevalence and characteristics of different types of headache in patients with Behcet's disease, a case-control study. *Headache.* 2008 Mar;48(3):424-9.

[17] Kidd D. The prevalence of headache in Behcet's syndrome. *Rheumatology* (Oxford). 2006 May;45(5):621-3.

[18] Aykutlu E, Baykan B, Akman-Demir G, Topcular B, Ertas M. Headache in Behcet's disease. *Cephalalgia.* 2006 Feb;26(2):180-6.

[19] Saip S, Siva A, Altintas A, Kiyat A, Seyahi E, Hamuryudan V, et al. Headache in Behcet's syndrome. *Headache.* 2005 Jul-Aug;45(7):911-9.

[20] Monastero R, Mannino M, Lopez G, Camarda C, Cannizzaro C, Camarda LK, et al. Prevalence of headache in patients with Behcet's disease without overt neurological involvement. *Cephalalgia.* 2003 Mar;23(2):105-8.

[21] Akman-Demir G, Tuzun E, Icoz S, Yesilot N, Yentur SP, Kurtuncu M, et al. Interleukin-6 in neuro-Behcet's disease: association with disease subsets and long-term outcome. *Cytokine.* 2008 Dec;44(3):373-6.

[22] Hirohata S, Isshi K, Oguchi H, Ohse T, Haraoka H, Takeuchi A, et al. Cerebrospinal fluid interleukin-6 in progressive Neuro-Behcet's syndrome. *Clin. Immunol. Immunopathol.* 1997 Jan;82(1):12-7.

[23] Hirohata S, Suda H, Hashimoto T. Low-dose weekly methotrexate for progressive neuropsychiatric manifestations in Behcet's disease. *J. Neurol. Sci.* 1998 Aug 14;159(2):181-5.

[24] Wang CR, Chuang CY, Chen CY. Anticardiolipin antibodies and interleukin-6 in cerebrospinal fluid and blood of Chinese patients with neuro-Behcet's syndrome. *Clin. Exp. Rheumatol.* 1992 Nov-Dec;10(6):599-602.

[25] Siva A, Fresko II. Behcet's Disease. *Curr. Treat Options Neurol.* 2000 Sep;2(5):435-48.

[26] O'Duffy JD, Robertson DM, Goldstein NP. Chlorambucil in the treatment of uveitis and meningoencephalitis of Behcet's disease. *Am. J. Med.* 1984 Jan;76(1):75-84.

[27] Kotter I, Gunaydin I, Batra M, Vonthein R, Stubiger N, Fierlbeck G, et al. CNS involvement occurs more frequently in patients with Behcet's disease under cyclosporin A (CSA) than under other medications-- results of a retrospective analysis of 117 cases. *Clin. Rheumatol.* 2006 Jul;25(4):482-6.

[28] Mitsui Y, Mitsui M, Urakami R, Kihara M, Takahashi M, Kusunoki S. Behcet disease presenting with neurological complications immediately after conversion from conventional cyclosporin A to microemulsion formulation. *Intern Med.* 2005 Feb;44(2):149-52.

[29] Kato Y, Numaga J, Kato S, Kaburaki T, Kawashima H, Fujino Y. Central nervous system symptoms in a population of Behcet's disease patients with refractory uveitis treated with cyclosporine A. *Clin. Experiment Ophthalmol.* 2001 Oct;29(5):335-6.

[30] Serkova NJ, Christians U, Benet LZ. Biochemical mechanisms of cyclosporine neurotoxicity. *Mol. Interv.* 2004 Apr;4(2):97-107.

[31] Kotake S, Higashi K, Yoshikawa K, Sasamoto Y, Okamoto T, Matsuda H. Central nervous system symptoms in patients with Behcet disease receiving cyclosporine therapy. *Ophthalmology.* 1999 Mar;106(3):586-9.

[32] Sarwar H, McGrath H, Jr., Espinoza LR. Successful treatment of long-standing neuro-Behcet's disease with infliximab. *J. Rheumatol.* 2005 Jan;32(1):181-3.

[33] Fujikawa K, Aratake K, Kawakami A, Aramaki T, Iwanaga N, Izumi Y, et al. Successful treatment of refractory neuro-Behcet's disease with infliximab: a case report to show its efficacy by magnetic resonance imaging, transcranial magnetic stimulation and cytokine profile. *Ann. Rheum. Dis.* 2007 Jan;66(1):136-7.

[34] Kikuchi H, Aramaki K, Hirohata S. Effect of infliximab in progressive neuro-Behcet's syndrome. *J. Neurol. Sci.* 2008 Sep 15;272(1-2):99-105.

[35] Licata G, Pinto A, Tuttolomondo A, Banco A, Ciccia F, Ferrante A, et al. Anti-tumour necrosis factor alpha monoclonal antibody therapy for recalcitrant cerebral vasculitis in a patient with Behcet's syndrome. *Ann. Rheum. Dis.* 2003 Mar;62(3):280-1.

[36] Ribi C, Sztajzel R, Delavelle J, Chizzolini C. Efficacy of TNF {alpha} blockade in cyclophosphamide resistant neuro-Behcet disease. *J. Neurol Neurosurg. Psychiatry.* 2005 Dec;76(12):1733-5.

[37] Pipitone N, Olivieri I, Padula A, D'Angelo S, Nigro A, Zuccoli G, et al. Infliximab for the treatment of Neuro-Behcet's disease: a case series and review of the literature. *Arthritis Rheum.* 2008 Feb 15;59(2):285-90.

[38] Sfikakis PP, Markomichelakis N, Alpsoy E, Assaad-Khalil S, Bodaghi B, Gul A, et al. Anti-TNF therapy in the management of Behcet's disease--review and basis for recommendations. *Rheumatology (Oxford)*. 2007 May;46(5):736-41.

[39] The American College of Rheumatology nomenclature and case definitions for neuropsychiatric lupus syndromes. *Arthritis Rheum*. 1999 Apr;42(4):599-608.

[40] Hanly JG, Urowitz MB, Sanchez-Guerrero J, Bae SC, Gordon C, Wallace DJ, et al. Neuropsychiatric events at the time of diagnosis of systemic lupus erythematosus: an international inception cohort study. *Arthritis Rheum*. 2007 Jan;56(1):265-73.

[41] Ainiala H, Loukkola J, Peltola J, Korpela M, Hietaharju A. The prevalence of neuropsychiatric syndromes in systemic lupus erythematosus. *Neurology*. 2001 Aug 14;57(3):496-500.

[42] Sanna G, Bertolaccini ML, Cuadrado MJ, Laing H, Khamashta MA, Mathieu A, et al. Neuropsychiatric manifestations in systemic lupus erythematosus: prevalence and association with antiphospholipid antibodies. *J. Rheumatol*. 2003 May;30(5):985-92.

[43] Brey RL, Holliday SL, Saklad AR, Navarrete MG, Hermosillo-Romo D, Stallworth CL, et al. Neuropsychiatric syndromes in lupus: prevalence using standardized definitions. *Neurology*. 2002 Apr 23;58(8):1214-20.

[44] Afeltra A, Garzia P, Mitterhofer AP, Vadacca M, Galluzzo S, Del Porto F, et al. Neuropsychiatric lupus syndromes: relationship with antiphospholipid antibodies. *Neurology*. 2003 Jul 8;61(1):108-10.

[45] Kasitanon N, Louthrenoo W, Piyasirisilp S, Sukitawu W, Wichainun R. Neuropsychiatric manifestations in Thai patients with systemic lupus erythematosus. *Asian Pac. J. Allergy Immunol*. 2002 Sep;20(3):179-85.

[46] Sibbitt WL, Jr., Brandt JR, Johnson CR, Maldonado ME, Patel SR, Ford CC, et al. The incidence and prevalence of neuropsychiatric syndromes in pediatric onset systemic lupus erythematosus. *J. Rheumatol*. 2002 Jul;29(7):1536-42.

[47] Harel L, Sandborg C, Lee T, von Scheven E. Neuropsychiatric manifestations in pediatric systemic lupus erythematosus and association with antiphospholipid antibodies. *J. Rheumatol*. 2006 Sep;33(9):1873-7.

[48] Bernatsky S, Clarke A, Gladman DD, Urowitz M, Fortin PR, Barr SG, et al. Mortality related to cerebrovascular disease in systemic lupus erythematosus. *Lupus*. 2006;15(12):835-9.

[49] Hanly JG. Neuropsychiatric lupus. *Curr. Rheumatol. Rep*. 2001 Jun;3(3):205-12.

[50] Jennekens FG, Kater L. The central nervous system in systemic lupus erythematosus. Part 2. Pathogenetic mechanisms of clinical syndromes: a literature investigation. *Rheumatology* (Oxford). 2002 Jun;41(6):619-30.

[51] Ainiala H, Hietaharju A, Dastidar P, Loukkola J, Lehtimaki T, Peltola J, et al. Increased serum matrix metalloproteinase 9 levels in systemic lupus erythematosus patients with neuropsychiatric manifestations and brain magnetic resonance imaging abnormalities. *Arthritis Rheum.* 2004 Mar;50(3):858-65.

[52] Zaccagni H, Fried J, Cornell J, Padilla P, Brey RL. Soluble adhesion molecule levels, neuropsychiatric lupus and lupus-related damage. *Front Biosci.* 2004 May 1;9:1654-9.

[53] Abbott NJ, Mendonca LL, Dolman DE. The blood-brain barrier in systemic lupus erythematosus. *Lupus.* 2003;12(12):908-15.

[54] Cervera R, Khamashta MA, Font J, Sebastiani GD, Gil A, Lavilla P, et al. Morbidity and mortality in systemic lupus erythematosus during a 10-year period: a comparison of early and late manifestations in a cohort of 1,000 patients. *Medicine* (Baltimore). 2003 Sep;82(5):299-308.

[55] McLaurin EY, Holliday SL, Williams P, Brey RL. Predictors of cognitive dysfunction in patients with systemic lupus erythematosus. *Neurology.* 2005 Jan 25;64(2):297-303.

[56] Hanly JG, Hong C, Smith S, Fisk JD. A prospective analysis of cognitive function and anticardiolipin antibodies in systemic lupus erythematosus. *Arthritis Rheum.* 1999 Apr;42(4):728-34.

[57] Menon S, Jameson-Shortall E, Newman SP, Hall-Craggs MR, Chinn R, Isenberg DA. A longitudinal study of anticardiolipin antibody levels and cognitive functioning in systemic lupus erythematosus. *Arthritis Rheum.* 1999 Apr;42(4):735-41.

[58] Yoshio T, Masuyama J, Ikeda M, Tamai K, Hachiya T, Emori T, et al. Quantification of antiribosomal P0 protein antibodies by ELISA with recombinant P0 fusion protein and their association with central nervous system disease in systemic lupus erythematosus. *J. Rheumatol.* 1995 Sep;22(9):1681-7.

[59] Schneebaum AB, Singleton JD, West SG, Blodgett JK, Allen LG, Cheronis JC, et al. Association of psychiatric manifestations with antibodies to ribosomal P proteins in systemic lupus erythematosus. *Am. J. Med.* 1991 Jan;90(1):54-62.

[60] Reichlin M. Ribosomal P antibodies and CNS lupus. *Lupus.* 2003;12(12):916-8.

[61] Lapteva L, Nowak M, Yarboro CH, Takada K, Roebuck-Spencer T, Weickert T, et al. Anti-N-methyl-D-aspartate receptor antibodies, cognitive dysfunction, and depression in systemic lupus erythematosus. *Arthritis Rheum.* 2006 Aug;54(8):2505-14.

[62] Omdal R, Brokstad K, Waterloo K, Koldingsnes W, Jonsson R, Mellgren SI. Neuropsychiatric disturbances in SLE are associated with antibodies against NMDA receptors. *Eur. J. Neurol.* 2005 May;12(5):392-8.

[63] Cunha BA. Infections in nonleukopenic compromised hosts (diabetes mellitus, SLE, steroids, and asplenia) in critical care. *Crit. Care Clin.* 1998 Apr;14(2):263-82.

[64] Mitchell SR, Nguyen PQ, Katz P. Increased risk of neisserial infections in systemic lupus erythematosus. *Semin. Arthritis Rheum.* 1990 Dec;20(3):174-84.

[65] Wilson JG, Ratnoff WD, Schur PH, Fearon DT. Decreased expression of the C3b/C4b receptor (CR1) and the C3d receptor (CR2) on B lymphocytes and of CR1 on neutrophils of patients with systemic lupus erythematosus. *Arthritis Rheum.* 1986 Jun;29(6):739-47.

[66] Noel V, Lortholary O, Casassus P, Cohen P, Genereau T, Andre MH, et al. Risk factors and prognostic influence of infection in a single cohort of 87 adults with systemic lupus erythematosus. *Ann. Rheum. Dis.* 2001 Dec;60(12):1141-4.

[67] Bermas BL, Petri M, Goldman D, Mittleman B, Miller MW, Stocks NI, et al. T helper cell dysfunction in systemic lupus erythematosus (SLE): relation to disease activity. *J. Clin. Immunol.* 1994 May;14(3):169-77.

[68] Pryor BD, Bologna SG, Kahl LE. Risk factors for serious infection during treatment with cyclophosphamide and high-dose corticosteroids for systemic lupus erythematosus. *Arthritis Rheum.* 1996 Sep;39(9):1475-82.

[69] Kraus A, Cabral AR, Sifuentes-Osornio J, Alarcon-Segovia D. Listeriosis in patients with connective tissue diseases. *J. Rheumatol.* 1994 Apr;21(4):635-8.

[70] Victorio-Navarra ST, Dy EE, Arroyo CG, Torralba TP. Tuberculosis among Filipino patients with systemic lupus erythematosus. *Semin. Arthritis Rheum.* 1996 Dec;26(3):628-34.

[71] Hung JJ, Ou LS, Lee WI, Huang JL. Central nervous system infections in patients with systemic lupus erythematosus. *J. Rheumatol.* 2005 Jan;32(1):40-3.

[72] Moris G, Garcia-Monco JC. The challenge of drug-induced aseptic meningitis. *Arch. Intern. Med.* 1999 Jun 14;159(11):1185-94.

[73] Berliner S, Weinberger A, Shoenfeld Y, Sandbank U, Hazaz B, Joshua H, et al. Ibuprofen may induce meningitis in (NZB X NZW)F1 mice. *Arthritis Rheum.* 1985 Jan;28(1):104-7.

[74] Shoenfeld Y, Livni E, Shaklai M, Pinkhas J. Sensitization to ibuprofen in systemic lupus erythematosus. *JAMA.* 1980 Aug 8;244(6):547-8.

[75] Kim JM, Kim KJ, Yoon HS, Kwok SK, Ju JH, Park KS, et al. Meningitis in Korean patients with systemic lupus erythematosus: analysis of demographics, clinical features and outcomes; experience from affiliated hospitals of the Catholic University of Korea. *Lupus.* 2011;20(5):531-6.

[76] Baizabal-Carvallo JF, Delgadillo-Marquez G, Estanol B, Garcia-Ramos G. Clinical characteristics and outcomes of the meningitides in systemic lupus erythematosus. *Eur. Neurol.* 2009;61(3):143-8.

[77] Gibson T, Myers AR. Nervous system involvement in systemic lupus erythematosus. *Ann. Rheum. Dis.* 1975 Oct;35(5):398-406.

[78] Canoso JJ, Cohen AS. Aspectic meningitis in systemic lupus erythematosus. Report of three cases. *Arthritis Rheum.* 1975 Jul-Aug;18(4):369-74.

[79] Johnson RT, Richardson EP. The neurological manifestations of systemic lupus erythematosus. *Medicine* (Baltimore). 1968 Jul;47(4):337-69.

[80] Andrianakos AA, Duffy J, Suzuki M, Sharp JT. Transverse myelopathy in systemic lupus erythematosus. Report of three cases and review of the literature. *Ann. Intern. Med.* 1975 Nov;83(5):616-24.

[81] Bertsias GK, Ioannidis JP, Aringer M, Bollen E, Bombardieri S, Bruce IN, et al. EULAR recommendations for the management of systemic lupus erythematosus with neuropsychiatric manifestations: report of a task force of the EULAR standing committee for clinical affairs. *Ann. Rheum. Dis.* 2010 Dec;69(12):2074-82.

[82] Bertsias GK, Boumpas DT. Pathogenesis, diagnosis and management of neuropsychiatric SLE manifestations. *Nat. Rev. Rheumatol.* 2010 Jun;6(6):358-67.

[83] Klippel JH. Indications for, and use of, cytotoxic agents in SLE. *Baillieres Clin. Rheumatol.* 1998 Aug;12(3):511-27.

[84] Jayne D, Passweg J, Marmont A, Farge D, Zhao X, Arnold R, et al. Autologous stem cell transplantation for systemic lupus erythematosus. *Lupus.* 2004;13(3):168-76.

[85] Traynor AE, Barr WG, Rosa RM, Rodriguez J, Oyama Y, Baker S, et al. Hematopoietic stem cell transplantation for severe and refractory lupus. Analysis after five years and fifteen patients. *Arthritis Rheum.* 2002 Nov;46(11):2917-23.

[86] Traynor AE, Schroeder J, Rosa RM, Cheng D, Stefka J, Mujais S, et al. Treatment of severe systemic lupus erythematosus with high-dose chemotherapy and haemopoietic stem-cell transplantation: a phase I study. *Lancet.* 2000 Aug 26;356(9231):701-7.

[87] Petri M, Brodsky R. High-dose cyclophosphamide and stem cell transplantation for refractory systemic lupus erythematosus. *JAMA.* 2006 Feb 1;295(5):559-60.

[88] Burt RK, Traynor A, Statkute L, Barr WG, Rosa R, Schroeder J, et al. Nonmyeloablative hematopoietic stem cell transplantation for systemic lupus erythematosus. *JAMA.* 2006 Feb 1;295(5):527-35.

[89] Ginzler EM, Dooley MA, Aranow C, Kim MY, Buyon J, Merrill JT, et al. Mycophenolate mofetil or intravenous cyclophosphamide for lupus nephritis. *N. Engl. J. Med.* 2005 Nov 24;353(21):2219-28.

[90] Leandro MJ, Edwards JC, Cambridge G, Ehrenstein MR, Isenberg DA. An open study of B lymphocyte depletion in systemic lupus erythematosus. *Arthritis Rheum.* 2002 Oct;46(10):2673-7.

[91] Looney RJ, Anolik JH, Campbell D, Felgar RE, Young F, Arend LJ, et al. B cell depletion as a novel treatment for systemic lupus erythematosus: a phase I/II dose-escalation trial of rituximab. *Arthritis Rheum.* 2004 Aug;50(8):2580-9.

[92] Sofat N, Malik O, Higgens CS. Neurological involvement in patients with rheumatic disease. *QJM.* 2006 Feb;99(2):69-79.

[93] Otsuka M, Fujiwara T, Kuwata Y, Yamada S, Ueki A. [A case of rheumatoid pachymeningitis]. *Rinsho Shinkeigaku.* 1997 Sep;37(9):834-40.

[94] Starosta MA, Brandwein SR. Clinical manifestations and treatment of rheumatoid pachymeningitis. *Neurology.* 2007 Mar 27;68(13):1079-80.

[95] Agildere AM, Tutar NU, Yucel E, Coskun M, Benli S, Aydin P. Pachymeningitis and optic neuritis in rheumatoid arthritis: MRI findings. *Br. J. Radiol.* 1999 Apr;72(856):404-7.

[96] Cellerini M, Gabbrielli S, Maddali Bongi S, Cammelli D. MRI of cerebral rheumatoid pachymeningitis: report of two cases with follow-up. *Neuroradiology.* 2001 Feb;43(2):147-50.

[97] Chowdhry V, Kumar N, Lachance DH, Salomao DR, Luthra HS. An unusual presentation of rheumatoid meningitis. *J. Neuroimaging.* 2005 Jul;15(3):286-8.

[98] Kurne A, Karabudak R, Karadag O, Yalcin-Cakmakli G, Karli-Oguz K, Yavuz K, et al. An unusual central nervous system involvement in rheumatoid arthritis: combination of pachymeningitis and cerebral vasculitis. *Rheumatol. Int.* 2009 Sep;29(11):1349-53.

[99] Ii Y, Kuzuhara S. Rheumatoid cranial pachymeningitis successfully treated with long-term corticosteroid. *Rheumatol. Int.* 2009 Mar;29(5):583-5.

[100] Yucel AE, Kart H, Aydin P, Agildere AM, Benli S, Altinors N, et al. Pachymeningitis and optic neuritis in rheumatoid arthritis: successful treatment with cyclophosphamide. *Clin. Rheumatol.* 2001;20(2):136-9.

[101] Markenson JA, McDougal JS, Tsairis P, Lockshin MD, Christian CL. Rheumatoid meningitis: a localized immune process. *Ann. Intern. Med.* 1979 May;90(5):786-9.

[102] Bathon JM, Moreland LW, DiBartolomeo AG. Inflammatory central nervous system involvement in rheumatoid arthritis. *Semin. Arthritis Rheum.* 1989 May;18(4):258-66.

[103] Hatano N, Behari S, Nagatani T, Kimura M, Ooka K, Saito K, et al. Idiopathic hypertrophic cranial pachymeningitis: clinicoradiological spectrum and therapeutic options. *Neurosurgery.* 1999 Dec;45(6):1336-42; discussion 42-4.

[104] Kupersmith MJ, Martin V, Heller G, Shah A, Mitnick HJ. Idiopathic hypertrophic pachymeningitis. *Neurology.* 2004 Mar 9;62(5):686-94.

[105] Martin N, Masson C, Henin D, Mompoint D, Marsault C, Nahum H. Hypertrophic cranial pachymeningitis: assessment with CT and MR imaging. *AJNR Am. J. Neuroradiol.* 1989 May-Jun;10(3):477-84.

[106] Shintani S, Shiigai T, Tsuruoka S. Hypertrophic cranial pachymeningitis causing progressive unilateral blindness: MR findings. *Clin. Neurol. Neurosurg.* 1993 Mar;95(1):65-70.

[107] Chou RC, Henson JW, Tian D, Hedley-Whyte ET, Reginato AM. Successful treatment of rheumatoid meningitis with cyclophosphamide but not infliximab. *Ann. Rheum. Dis.* 2006 Aug;65(8):1114-6.

[108] Claassen J, Dwyer E, Maybaum S, Elkind MS. Rheumatoid leptomeningitis after heart transplantation. *Neurology.* 2006 Mar 28;66(6):948-9.

[109] Jones SE, Belsley NA, McLoud TC, Mullins ME. Rheumatoid meningitis: radiologic and pathologic correlation. *AJR Am. J. Roentgenol.* 2006 Apr;186(4):1181-3.

[110] Kato T, Hoshi K, Sekijima Y, Matsuda M, Hashimoto T, Otani M, et al. Rheumatoid meningitis: an autopsy report and review of the literature. *Clin. Rheumatol.* 2003 Dec;22(6):475-80.

[111] Matsushima M, Yaguchi H, Niino M, Akimoto-Tsuji S, Yabe I, Onishi K, et al. MRI and pathological findings of rheumatoid meningitis. *J. Clin. Neurosci.* 2010 Jan;17(1):129-32.

[112] Matsuura D, Ohshita T, Nagano Y, Ohtsuki T, Kohriyama T, Matsumoto M. [Case of rheumatoid meningitis: findings on diffusion-weighted image versus FLAIR image]. *Rinsho Shinkeigaku.* 2008 Mar;48(3): 191-5.

[113] Shimada K, Matsui T, Kawakami M, Hayakawa H, Futami H, Michishita K, et al. Diffuse chronic leptomeningitis with seropositive rheumatoid arthritis: report of a case successfully treated as rheumatoid leptomeningitis. *Mod. Rheumatol.* 2009;19(5):556-62.

[114] Yaguchi M, Yaguchi H. Unilateral supratentorial lesion due to rheumatoid meningitis on MRI. *Intern. Med.* 2008;47(21):1947-8.

[115] Ahmed M, Luggen M, Herman JII, Weiss KL, Decourten-Myers G, Quinlan JG, et al. Hypertrophic pachymeningitis in rheumatoid arthritis after adalimumab administration. *J. Rheumatol.* 2006 Nov;33(11): 2344-6.

[116] Mauch E, Volk C, Kratzsch G, Krapf H, Kornhuber HH, Laufen H, et al. Neurological and neuropsychiatric dysfunction in primary Sjogren's syndrome. *Acta Neurol. Scand.* 1994 Jan;89(1):31-5.

[117] Govoni M, Bajocchi G, Rizzo N, Tola MR, Caniatti L, Tugnoli V, et al. Neurological involvement in primary Sjogren's syndrome: clinical and instrumental evaluation in a cohort of Italian patients. *Clin. Rheumatol.* 1999;18(4):299-303.

[118] Alexander EL, Provost TT, Stevens MB, Alexander GE. Neurologic complications of primary Sjogren's syndrome. *Medicine* (Baltimore). 1982 Jul;61(4):247-57.

[119] Binder A, Snaith ML, Isenberg D. Sjogren's syndrome: a study of its neurological complications. *Br. J. Rheumatol.* 1988 Aug;27(4):275-80.

[120] Alexander EL, Alexander GE. Aseptic meningoencephalitis in primary Sjogren's syndrome. *Neurology.* 1983 May;33(5):593-8.

[121] Alexander GE, Provost TT, Stevens MB, Alexander EL. Sjogren syndrome: central nervous system manifestations. *Neurology.* 1981 Nov;31(11):1391-6.

[122] De Backer H, Dehaene I. Central nervous system disease in primary Sjogren's syndrome. *Acta Neurol. Belg.* 1995;95(3):142-6.

[123] Rossi R, Valeria Saddi M. Subacute aseptic meningitis as neurological manifestation of primary Sjogren's syndrome. *Clin. Neurol. Neurosurg.* 2006 Oct;108(7):688-91.

[124] Caselli RJ, Scheithauer BW, O'Duffy JD, Peterson GC, Westmoreland BF, Davenport PA. Chronic inflammatory meningoencephalitis should not be mistaken for Alzheimer's disease. *Mayo Clin. Proc.* 1993 Sep;68(9):846-53.

[125] Alexander EL, Hirsch TJ, Arnett FC, Provost TT, Stevens MB. Ro(SSA) and La(SSB) antibodies in the clinical spectrum of Sjogren's syndrome. *J. Rheumatol.* 1982 Mar-Apr;9(2):239-46.

[126] Delalande S, de Seze J, Fauchais AL, Hachulla E, Stojkovic T, Ferriby D, et al. Neurologic manifestations in primary Sjogren syndrome: a study of 82 patients. *Medicine* (Baltimore). 2004 Sep;83(5):280-91.

[127] Hoshina T, Yamaguchi Y, Ohga S, Kira R, Ishimura M, Takada H, et al. Sjogren's syndrome-associated meningoencephalomyelitis: cerebrospinal fluid cytokine levels and therapeutic utility of tacrolimus. *J. Neurol. Sci.* 2008 Apr 15;267(1-2):182-6.

[128] Ishida K, Uchihara T, Mizusawa H. Recurrent aseptic meningitis: a new CSF complication of Sjogren's syndrome. *J. Neurol.* 2007 Jun;254(6):806-7.

[129] Vincent TL, Richardson MP, Mackworth-Young CG, Hawke SH, Venables PJ. Sjogren's syndrome-associated myelopathy: response to immunosuppressive treatment. *Am. J. Med.* 2003 Feb 1;114(2):145-8.

[130] Hawley RJ, Hendricks WT. Treatment of Sjogren syndrome myelopathy with azathioprine and steroids. *Arch. Neurol.* 2002 May;59(5):875; author reply 6.

[131] Williams CS, Butler E, Roman GC. Treatment of myelopathy in Sjogren syndrome with a combination of prednisone and cyclophosphamide. *Arch. Neurol.* 2001 May;58(5):815-9.

[132] Rogers SJ, Williams CS, Roman GC. Myelopathy in Sjogren's syndrome: role of nonsteroidal immunosuppressants. *Drugs.* 2004;64(2):123-32.

[133] Yamout B, El-Hajj T, Barada W, Uthman I. Successful treatment of refractory neuroSjogren with Rituximab. *Lupus.* 2007;16(7):521-3.

[134] Ozgocmen S, Gur A. Treatment of central nervous system involvement associated with primary Sjogren's syndrome. *Curr. Pharm. Des.* 2008;14(13):1270-3.

[135] Gerraty RP, McKelvie PA, Byrne E. Aseptic meningoencephalitis in primary Sjogren's syndrome. Response to plasmapheresis and absence of CNS vasculitis at autopsy. *Acta Neurol. Scand.* 1993 Oct;88(4):309-11.

[136] Moller DR, Chen ES. What causes sarcoidosis? *Curr. Opin. Pulm. Med.* 2002 Sep;8(5):429-34.

[137] Iannuzzi MC, Rybicki BA, Teirstein AS. Sarcoidosis. *N. Engl. J. Med.* 2007 Nov 22;357(21):2153-65.

[138] Rybicki BA, Iannuzzi MC. Epidemiology of sarcoidosis: recent advances and future prospects. *Semin. Respir. Crit. Care Med.* 2007 Feb;28(1):22-35.

[139] Oksanen V. Neurosarcoidosis: clinical presentations and course in 50 patients. *Acta Neurol. Scand.* 1986 Mar;73(3):283-90.

[140] Chen RC, McLeod JG. Neurological complications of sarcoidosis. *Clin. Exp. Neurol.* 1989;26:99-112.

[141] Stern BJ, Krumholz A, Johns C, Scott P, Nissim J. Sarcoidosis and its neurological manifestations. *Arch. Neurol.* 1985 Sep;42(9):909-17.

[142] James DG, Sharma OP. Neurosarcoidosis. *Proc. R. Soc. Med.* 1967 Nov 1;60(11 Part 1):1169-70.

[143] Iwai K, Tachibana T, Takemura T, Matsui Y, Kitaichi M, Kawabata Y. Pathological studies on sarcoidosis autopsy. I. Epidemiological features of 320 cases in Japan. *Acta Pathol. Jpn.* 1993 Jul-Aug;43(7-8):372-6.

[144] Zajicek JP, Scolding NJ, Foster O, Rovaris M, Evanson J, Moseley IF, et al. Central nervous system sarcoidosis--diagnosis and management. *QJM.* 1999 Feb;92(2):103-17.

[145] Marangoni S, Argentiero V, Tavolato B. Neurosarcoidosis. Clinical description of 7 cases with a proposal for a new diagnostic strategy. *J. Neurol.* 2006 Apr;253(4):488-95.

[146] Pawate S, Moses H, Sriram S. Presentations and outcomes of neurosarcoidosis: a study of 54 cases. *QJM.* 2009 Jul;102(7):449-60.

[147] Joseph FG, Scolding NJ. Neurosarcoidosis: a study of 30 new cases. *J. Neurol. Neurosurg. Psychiatry.* 2009 Mar;80(3):297-304.

[148] Sharma OP. Neurosarcoidosis: a personal perspective based on the study of 37 patients. *Chest.* 1997 Jul;112(1):220-8.

[149] Nowak DA, Widenka DC. Neurosarcoidosis: a review of its intracranial manifestation. *J. Neurol.* 2001 May;248(5):363-72.

[150] Oksanen V, Fyhrquist F, Somer H, Gronhagen-Riska C. Angiotensin converting enzyme in cerebrospinal fluid: a new assay. *Neurology*. 1985 Aug;35(8):1220-3.

[151] Tahmoush AJ, Amir MS, Connor WW, Farry JK, Didato S, Ulhoa-Cintra A, et al. CSF-ACE activity in probable CNS neurosarcoidosis. *Sarcoidosis Vasc. Diffuse Lung Dis*. 2002 Oct;19(3):191-7.

[152] Dale JC, O'Brien JF. Determination of angiotensin-converting enzyme levels in cerebrospinal fluid is not a useful test for the diagnosis of neurosarcoidosis. *Mayo Clin. Proc*. 1999 May;74(5):535.

[153] Joseph FG, Scolding NJ. Sarcoidosis of the nervous system. *Pract. Neurol*. 2007 Aug;7(4):234-44.

[154] Miller DH, Kendall BE, Barter S, Johnson G, MacManus DG, Logsdail SJ, et al. Magnetic resonance imaging in central nervous system sarcoidosis. *Neurology*. 1988 Mar;38(3):378-83.

[155] Khaw KT, Manji H, Britton J, Schon F. Neurosarcoidosis--demonstration of meningeal disease by gadolinium enhanced magnetic resonance imaging. *J. Neurol. Neurosurg. Psychiatry*. 1991 Jun;54(6):499-502.

[156] Lower EE, Baughman RP. The use of low dose methotrexate in refractory sarcoidosis. *Am. J. Med. Sci*. 1990 Mar;299(3):153-7.

[157] Soriano FG, Caramelli P, Nitrini R, Rocha AS. Neurosarcoidosis: therapeutic success with methotrexate. *Postgrad. Med. J*. 1990 Feb;66(772):142-3.

[158] Stern BJ, Schonfeld SA, Sewell C, Krumholz A, Scott P, Belendiuk G. The treatment of neurosarcoidosis with cyclosporine. *Arch Neurol*. 1992 Oct;49(10):1065-72.

[159] Androdias G, Maillet D, Marignier R, Pinede L, Confavreux C, Broussolle C, et al. Mycophenolate mofetil may be effective in CNS sarcoidosis but not in sarcoid myopathy. *Neurology*. 2011 Mar 29;76(13):1168-72.

[160] Bullmann C, Faust M, Hoffmann A, Heppner C, Jockenhovel F, Muller-Wieland D, et al. Five cases with central diabetes insipidus and hypogonadism as first presentation of neurosarcoidosis. *Eur. J. Endocrinol*. 2000 Apr;142(4):365-72.

[161] Baughman RP, Strohofer SA, Buchsbaum J, Lower EE. Release of tumor necrosis factor by alveolar macrophages of patients with sarcoidosis. *J. Lab. Clin. Med*. 1990 Jan;115(1):36-42.

[162] Zheng L, Teschler H, Guzman J, Hubner K, Striz I, Costabel U. Alveolar macrophage TNF-alpha release and BAL cell phenotypes in sarcoidosis. *Am. J. Respir. Crit. Care Med.* 1995 Sep;152(3):1061-6.

[163] Doty JD, Mazur JE, Judson MA. Treatment of sarcoidosis with infliximab. *Chest.* 2005 Mar;127(3):1064-71.

[164] Sollberger M, Fluri F, Baumann T, Sonnet S, Tamm M, Steck AJ, et al. Successful treatment of steroid-refractory neurosarcoidosis with infliximab. *J. Neurol.* 2004 Jun;251(6):760-1.

[165] Carter JD, Valeriano J, Vasey FB, Bognar B. Refractory neurosarcoidosis: a dramatic response to infliximab. *Am. J. Med.* 2004 Aug 15;117(4):277-9.

[166] Morcos Z. Refractory neurosarcoidosis responding to infliximab. *Neurology.* 2003 Apr 8;60(7):1220-1; author reply -1.

[167] Santos E, Shaunak S, Renowden S, Scolding NJ. Treatment of refractory neurosarcoidosis with Infliximab. *J. Neurol. Neurosurg. Psychiatry.* 2010 Mar;81(3):241-6.

[168] Pettersen JA, Zochodne DW, Bell RB, Martin L, Hill MD. Refractory neurosarcoidosis responding to infliximab. *Neurology.* 2002 Nov 26;59(10):1660-1.

[169] Fujita Y, Fujii T, Nakashima R, Tanaka M, Mimori T. Aseptic meningitis in mixed connective tissue disease: cytokine and anti-U1RNP antibodies in cerebrospinal fluids from two different cases. *Mod. Rheumatol.* 2008;18(2):184-8.

[170] Ginsberg L, Kidd D. Chronic and recurrent meningitis. *Pract. Neurol.* 2008 Dec;8(6):348-61.

[171] Zhang W, Zhou G, Shi Q, Zhang X, Zeng XF, Zhang FC. Clinical analysis of nervous system involvement in ANCA-associated systemic vasculitides. *Clin. Exp. Rheumatol.* 2009 Jan-Feb;27(1 Suppl 52):S65-9.

[172] Birnbaum J, Hellmann DB. Primary angiitis of the central nervous system. *Arch. Neurol.* 2009 Jun;66(6):704-9.

[173] Pohl D, Rostasy K, Reiber H, Hanefeld F. CSF characteristics in early-onset multiple sclerosis. *Neurology.* 2004 Nov 23;63(10):1966-7.

[174] Wingerchuk DM, Lennon VA, Pittock SJ, Lucchinetti CF, Weinshenker BG. Revised diagnostic criteria for neuromyelitis optica. *Neurology.* 2006 May 23;66(10):1485-9.

[175] Nagashima T, Maguchi S, Terayama Y, Horimoto M, Nemoto M, Nunomura M, et al. P-ANCA-positive Wegener's granulomatosis presenting with hypertrophic pachymeningitis and multiple cranial

neuropathies: case report and review of literature. *Neuropathology*. 2000 Mar;20(1):23-30.

[176] Furukawa Y, Matsumoto Y, Yamada M. Hypertrophic pachymeningitis as an initial and cardinal manifestation of microscopic polyangiitis. *Neurology*. 2004 Nov 9;63(9):1722-4.

[177] Kono H, Inokuma S, Nakayama H, Yamazaki J. Pachymeningitis in microscopic polyangiitis (MPA): a case report and a review of central nervous system involvement in MPA. *Clin. Exp. Rheumatol.* 2000 May-Jun;18(3):397-400.

[178] Takahashi K, Kobayashi S, Okada K, Yamaguchi S. Pachymeningitis with a perinuclear antineutrophil cytoplasmic antibody: response to pulse steroid. *Neurology*. 1998 Apr;50(4):1190-1.

In: Meningitis
Editor: Anthony L. Shrader

ISBN: 978-1-63117-136-9
© 2014 Nova Science Publishers, Inc.

Chapter 6

Sepsis and Bacteremia, Meningitis, Influenza Infection and Infectious Diarrhea

Elif Doyuk-Kartal[1] and Hakan Erdem[2]
[1]Osmangazi University, School of Medicine, Department of Infectious
Diseases and Clinical Microbiology, Eskisehir, Turkey
[2]GATA Haydarpasa Hospital,
Department of Infectious Diseases and Clinical Microbiology,
Istanbul, Turkey

Sepsis and Bacteremia

Mortality in adults is frequently associated with sepsis, which becomes more common with advancing age. The mortality rate is five out of 100,000 in patients younger than 65 and increases to up to 26 for those over the age of 85 [1]. In elderly patients treated in the emergency department (ED), sepsis and bacteremia were reported to be involved in 18% of cases [2]. In another report, positive blood cultures were obtained for approximately 10 % of the elderly patients in EDs. Escherichia coli was the most commonly isolated pathogen (29%), and the urinary tract was the most common source of infection (44 %) [3].

Several factors have been suggested to facilitate the acquisition of urinary tract infections in this patient population. Decreasing physiological and immunological functions, malnutrition, the presence of comorbid diseases, the use of urinary catheters, in-hospital treatment, and residing in long-term care facilities are well-known predisposing conditions [4, 5].

Multiple changes occur in the immune system with advancing age. Although the basal components of natural immunity functions regularly, the effectiveness of the acquired immune system decreases. The functions of antigen presenting cells, natural killer cells, and neutrophils are preserved, albeit with minimal losses. The other components of the humoral immunity and B cell responses are also affected [6]. Consequently, these changes allow for the dissemination of infectious diseases in the elderly population.

The signs and symptoms of sepsis are generally atypical in older individuals, and these results in delays in the diagnosis and treatment of the patients. Emergency physicians detect the presence of bacteremia in less than two-thirds of cases [3]. Fever, a major symptom in severe infectious disease patients, can be undetectable in patients with blood stream infections. Afebrile bacteremia was found to be more common among elderly patients compared to young patients [7]. In a study evaluating elderly patients with bacteremia, fever was not observed in 15-30 % of the cases. Furthermore, fewer than 20% of the bacteremic elderly patients complained of fever preceding ED admission [3].

The physician should be aware that elderly patients with sepsis present different symptoms than younger patients. In the elderly patient population, nonspecific indicators of infections, such as fatigue, weakness, and urinary incontinence, and mental changes, such as confusion or delirium, and even strokes, are frequently recorded [8, 9]. The clinical presentation at the beginning of sepsis may vary, but patients can rapidly progress to severe sepsis in a very short time. Tolerance to stress decreases in tissues with aging, and multiple organ failures can be observed in elderly patients [5]. In a large-scale study, tachypnea and mental changes were more frequent, and tachycardia and hypoxemia were less common in septic patients over the age of 75 compared to septic patients younger than 75 [10]. In the identification of the elderly with bacteremia, the place and efficacy of either the laboratory or physical examination methods can also be limited [4]. In another large study, fever, respiratory symptoms, urinary complaints, vital signs, and white blood cell (WBC) counts were evaluated as diagnostic markers of bacteremia in elderly patients. In that study, altered mental status, vomiting, and WBC bands exceeding 6 % were reported to be independent risk factors [3]. Chassagne et al. noted that a rapid onset of disease, fever, an identified source of infection, and an altered general state were independent risk factors for bacteremia in elderly patients [11]. The WBC count is not a perfect indicator in bacteremia because 20-45 % of patients are reported to have WBC counts within the normal range [8]. Furthermore, no guideline exists for the management of elderly patients with sepsis. A two-step protocol can be used in the ED. The first step considers the clinical picture, which cannot be explained with a noninfectious disorder, with the presence of at least two of the systemic inflammatory response criteria within the first six hours of admission. The second step involves the confirmation of the situation in favor of sepsis by the treating clinician. This two-step approach is known to detect serious infections with 98 % specificity but with only 11 % sensitivity [12].

An identified source of infection is frequently not known in sepsis or bacteremia. In 10-30 % of blood culture-positive elderly patients, the site of infectious disease could not be determined [8]. The main sources of community-acquired bacteremia are the urinary, gastrointestinal, and respiratory systems [8, 9]. The urinary system was reported to be the source of bacteremia in 25-55 % of patients in the ED, followed by the respiratory tract in 10-34 %; in 10-31% of patients, the source was undetermined. In addition, an intra-abdominal source was reported for approximately 9-20 % of the patients, and skin and catheter-related sources were reported for 9 % of the patients [3, 8, 13]. In the elderly population, predisposing factors and bacteremia etiologies were found to be closely interrelated. While the skin is the probable source of infection in a patient with an intravascular device, the urinary system is the likely source when the patient has a coexisting urinary catheter. Furthermore, altered consciousness can compromise the usual defenses that protect the lower airways, including glottic closure, the cough reflex, and other clearing mechanisms that have been implicated in bacteremia originating from the respiratory system [8, 9].

The causative agents appear to vary according to the source of the bacteremia. In urinary system-related bacteremia, Gram-negative enteric bacteria and enterococci are most common. In bacteremia associated with the biliary system, Gram-negative enteric bacteria and anaerobes are frequently identified as the causative agents. Haemophilus influenzae, Streptococcus pneumoniae, Group B streptococci, and Gram-negative enteric bacteria have been defined as the major agents invading the blood stream in cases associated with the respiratory system. When the infection originates from the skin, Staphylococcus aureus, Staphylococcus epidermidis, Gram-negative enteric bacteria and anaerobes can be the infecting agents [8]. Normally, 70 % of the infecting agents are Gram-negative, and 25 % are Gram-positive organisms; Escherichia coli is the most commonly isolated microorganism. Anaerobes are detected in less than 10 % of the cases, and a polymicrobial etiology is observed in 5-17 % of all patients [3, 13].

Although various ICU treatment strategies have been applied to the management of sepsis, discrepancies existed in managing young and elderly patient populations. For example, in the resuscitation of elderly patients with sepsis in the ED, interventions should take into consideration the risks of volume load, arrhythmias due to drug infusions, and complications during the placement of central catheters [4, 14, 15]. The treatment strategies for elderly patients with sepsis and septic shock should be more aggressive compared to the treatment of younger adults. Antibiotics should be started within one hour of diagnosing sepsis or septic shock [8, 15]. The antibiotic should be selected based on the source of infection, patient characteristics, and the local resistance patterns. Multidrug-resistant flora may be a concern in the elderly population due to frequent use of antimicrobials, and mortality is known to increase with inefficient antibiotic coverage. Thus, the spectrum of antibiotics for the elderly patients should be relatively broad [15, 16]. Decreases in renal function, body mass and liver blood flow may seriously alter the pharmacokinetics and pharmacodynamics of the antimicrobial drugs. Hence, careful therapeutic drug monitoring is indicated for this patient population [15].

The use of aggressive resuscitation in elderly patients in EDs is a very critical event. In one study, significant improvements in the survival of septic shock patients with a mean age of 65.7 with early aggressive resuscitation in the ED were reported [17]. Adequate restoration of organ perfusion, as measured by central venous pressure and urinary output, should be initiated rapidly and maintained. The primary goals are to restore effective circulatory volume, to maintain mean arterial pressure, and to provide sufficient oxygen to the extremities. Intravenous fluid replacement should be started to restore the effective circulatory volume. During the initial resuscitation, intravenous fluids (crystalloids and colloids) can be used freely. The goal is to maintain a central venous pressure of approximately 8-12 mmHg (>12 mmHg in intubated patients). If tissue perfusion cannot be restored despite adequate fluid replacement, critically important hypotension vasopressors, such as epinephrine or dopamine, can be administered. [4].

The oxygenation of the extremities can be monitored by measuring the central venous oxygen saturation (CVOS). CVOS less than 70 %demonstrates inadequate oxygenation, and transfusion is indicated under this circumstance

[4, 15, 18]. The threshold of transfusion is determined as 7 mg/dL, excluding cases with a distorted coronary anatomy and elderly patients with a history of myocardial infarction [4]. Intravenous corticosteroid use has also been recommended for the elderly population with septic shock requiring persistent vasopressors [15, 19]. A low dose or replacement dose of steroids has been shown to improve hemodynamics and decrease mortality [19]. However, other reports indicate that corticosteroids are not helpful in decreasing mortality and may instead increase super infections [20].

The PROWESS study evaluated the use of activated protein C and found that this molecule contributed to both short- and long-term survival in patients with septic shock over the age of 75 [21]. When the appropriate criteria were established, drotrecoginalfa has been safely used in elderly patients. Furthermore, the use of drotrecoginalfa has been increasingly reported in EDs, and if rational protocols or guidelines can be established, the use of this molecule can become more widespread in EDs [22, 23]. However, it is quite expensive [19].

Diabetes mellitus is a frequent disease in the elderly. Close glycemic control has improved the survival rates in critical cases. The goal is to maintain blood glucose levels approximately 80-110 mg/dl with the use of insulin [24]. However, diabetic patients should be closely monitored due to their risk of developing hypoglycemia [19, 25]. In addition, for hemodynamically unstable severe sepsis patients, performing early and sufficient dialysis is known to improve survival [4]. Furthermore, mechanical ventilation is frequently indicated in elderly patients with severe sepsis, and mortality is reported to decrease with the administration of a low tidal volume in patients below the age of 70 [4].

Meningitis

Meningitis in the elderly population has higher case-fatality rates compared to younger adults, as in all serious infections. Moreover, mortality is reported to increase as age increases [26, 27]. In a study that evaluated 493 adult bacterial meningitis cases over a 27-year period, the risk factors for mortality were defined as mental alterations upon admission, age over 60, and the detection of seizures in the first 24 hours [28]. In our study, the risk factors for fatality in the elderly population were increasing age, the presence of stupor, sepsis, and inappropriate antibiotic administration [27]. The morbidity and mortality rates of elderly patients with serious neurological disorders were reported to be as high as 50 % [29]. The leading pathogens causing meningitis were S. pneumoniae, Listeria monocytogenes, Gram-negative bacilli and Streptococcus agalactiae. Neisseria meningitidis, H. influenza and viral agents appear to be infrequent in the elderly population [27, 30-32].

Due to the silent nature of the disease in the elderly, the diagnosis of meningitis in this age group is difficult. The physician in the ED should be aware of the possibility of meningitis, especially for patients at the extreme ages of life [26]. Fever is not as common in elderly patients as it is in younger adults. Neck stiffness does not always occur in elderly patients with meningitis. It should be considered that there may other physical limitations to the neck movements of elderly patients, and a degree of nuchal rigidity may frequently be present in the absence of central nervous system infections. In the absence of other accompanying neurological manifestations, this finding can mistakenly be interpreted as osteoarthritis. Thus, the presence of probable meningitis should be taken into consideration [29, 33]. Although laboratory findings do not differ considerably in the elderly compared to younger adults, altered mentation, respiratory symptoms and seizures are more common in this patient population [33, 34].

The diagnosis should be established as soon as possible in elderly patients with bacterial meningitis and should be followed by rational antibiotic treatment. Unfortunately, the start of antibiotics can be delayed in the ED due to pending cranial tomography and cerebrospinal fluid analysis results [35]. Consequently, antibiotics should be started immediately when the initial evaluations suggest the probability of acute bacterial meningitis [26, 29, 35]. The antibiotics should cover S. pneumoniae, Listeria monocytogenes, and Gram-negative bacilli. L. monocytogenesis is not susceptible to cephalosporins. The rational antibiotic regime includes the combination of ampicillin or penicillin with ceftriaxone or cefotaxime in patients over the age of 50 [27]. In a given community where penicillin-resistant S.pneumoniae is known to be endemic, the use of vancomycin as a part of the combination is recommended [9, 29, 34, 36].

Although there have been ongoing debates regarding the use of dexamethasone in the management of acute bacterial meningitis, mortality was reported to decline with steroid use in various studies [37, 38]. Thus, the indications for dexamethasone use are detailed in the current guidelines [36, 39]. Dexamethasone was found to be the most useful in pneumococcal meningitis patients and in those with a moderate to severe Glasgow's coma scale [38]. Dexamethasone also appears to be useful in treating comatose patients, patients with increased cerebrospinal fluid pressure, and patients in whom Gram-positive diplococcic have been observed by Gram stain [26]. The optimal timing of dexamethasone administration is crucial, and the drug should be given 15-20 minutes preceding or concordant with the first dose of the antibiotics. If the antibiotics have already been administered, then steroids should not be allowed. The usual dose of dexamethasone is 10 mg every six hours for 2-4 days [36, 39, 40].

Influenza Infection

Patients over the age of 65are more vulnerable to influenza infection due to immunological deficits that appear with advanced age [41]. During the 2009 H1N1 pandemic, the attack rate was the highest among the children and young adults, while the elders over the age of 60 made up less than 10 % of the cases. However, advanced age was a risk factor for poor prognosis. The highest mortality rate was observed in patients over the age of 65 [42-44]. Accordingly, the hospitalization rate of people over 65 was remarkably high, and the mortality among the hospitalized patients in this subgroup was recorded to be approximately 90 % [45, 46]. Moreover, complications due to underlying conditions in the pulmonary and cardiovascular systems are common in elderly individuals [47, 48]. For this reason, influenza infection is a noteworthy burden in both the community and the healthcare setting [41].

The diagnosis of influenza can be difficult in the elderly population. A substantial portion of the patients arrive at medical centers with severe respiratory symptoms in the absence of classical complaints of influenza, such as fever, cough, fatigue, myalgia, and rhinorrhea [42]. Laboratory diagnosis of influenza is also difficult, and rapid tests have been developed for the detection of influenza antigens in the respiratory secretions. Although the results of these tests are available within one hour, the sensitivity range is between 50-90% [49, 50].

Older antiviral drugs, such as amantadine and rimantadine, are not recommended due to resistance and frequent side effects in the elderly population [51]. Other antiviral drugs, such as oseltamivir and zanamivir, have been used to treat influenza infection. Currently, resistance to these antivirals is rare [42]. Oseltamivir is given twice daily for five days in 75 mg doses. Its dose can be adjusted according to creatinine clearance. Zanamivir can result in bronchospasm attacks in elderly patients with underlying respiratory problems [51].

Influenza vaccination is the cornerstone in the control of this disease. Currently, only the inactivated trivalent influenza vaccine, which has been on the market for more than 60 years, is available and recommended for the elderly [41]. As a rule, elderly individuals should be immunized annually [48]. Although the efficacy of the vaccine was found to be lower in the elderly compared to younger adults [52], it decreases hospital visits by one-third and the death rate by 50 % [41]. Immunization programs are successfully conducted in EDs, and physicians in the ED should suggest the influenza vaccination to elderly patients who are at the ED for any reason. The vaccine should not be given in case of egg allergy. Guillian-Barre syndrome and moderate to severe febrile illness are the relative contraindications [41, 48].

Infectious Diarrhea

Diarrhea is an important cause of morbidity and mortality in the elderly. The disease is of particular importance for the residents of nursing homes and has a considerable mortality as high as 17.5 % in this subset of the population [53, 54]. In a study, age over 70 was reported to be a risk factor for poor prognosis in Clostridium difficile-associated diarrhea [55]. Various well-known factors, such as achlorhydria, decreasing intestinal motility, frequent antibiotic use, alterations in the immune system, comorbid diseases, and being a resident of a nursing home, have been reported as risk factors for infectious diarrhea in the elderly population [9, 56]. Enterohemorrhagic E. coli (O157:H7), Clostridium difficile, noroviruses, and rotaviruses can result in outbreaks in nursing homes. The elderly were reported to be more vulnerable to norovirus and rotavirus outbreaks in the winter months at these facilities [9]. C. difficile is one of the leading causes of sporadic diarrhea and is reported to be the most common causative agents of diarrhea in the elderly population in England [57]. This microorganism has the potential to cause pseudomembranous colitis; thus, mortality increases with age and doubles after the seventh decade [55, 58]. The symptoms of infectious diarrhea in the elderly can be atypical and can be difficult to discriminate from fecal incontinence.

Certain agents may result in systemic disease [57]. Invasive Salmonella infections are also frequent in the elderly, and the mortality rate for these infections was reported to be 45 % for individuals over the age of 45 [59]. The secondary attack rate of the diarrheas related to Cryptosporidium is high and has been reported to be potentially more serious in the elderly [60]. Although Candida can be the causative agent of antibiotic-associated colitis in this patient population, the isolation of fungi from the stool does not directly suggest infection. Rather, it is primarily an indicator of colonization [61].

These infections in the elderly population may be intolerable. The maintenance and replacement of fluid and electrolytes is extremely important. Antibiotics are indicated in the management of infectious diarrhea resulting from particular microorganisms, such as Shigella spp, Campylobacter jejuni, invasive E. coli, Vibrio parahaemolyticus and Yersinia enterocolitica. Some authors recommend the use of antibiotics in the management of uncomplicated Salmonella gastroenteritis in individuals over the age of 50. When a bacterial etiology is suspected, quinolones are the rational choice for the empirical management of acute gastroenteritis. Antiperistaltic drugs are not recommended due to their adverse effects [9, 62].

References

[1]   Angus DC, Linde-Zwirble WT, Lidicker J, Clermont G, Carcillo J, Pinsky MR. Epidemiology of severe sepsis in the United States: analysis of incidence, outcome, and associated costs of care. *Crit. Care Med.* 2001; 29(7): 1303-10.

[2]   Marco CA, Schoenfeld CN, Hansen KN, Hexter DA, Stearns DA, Kelen GD. Fever in geriatric emergency patients: clinical features associated with serious illness. *Ann. Emerg. Med.* 1995; 26(1): 18-24.

[3]   Fontanarosa PB, Kaeberlein FJ, Gerson LW, Thomson RB. Difficulty in predicting bacteremia in elderly emergency patients. *Ann. Emerg. Med.* 1992; 21(7): 842-8.

[4]   Girard TD, Opal SM, Ely EW. Insights into severe sepsis in older patients: from epidemiology to evidence-based management. *Clin. Infect. Dis.* 2005; 40(5): 719-27.

[5]   De Gaudio AR, Rinaldi S, Chelazzi C, Borracci T. Pathophysiology of sepsis in the elderly: clinical impact and therapeutic considerations. *Curr. Drug Targets* 2009; 10(1): 60-70.

[6]   Grubeck-Loebenstein B, Wick G. The aging of the immune system. *Adv. Immunol.* 2002; 80: 243-84.

[7]   Gleckman R, Hibert D. Afebrile bacteremia. A phenomenon in geriatric patients. *JAMA* 1982; 248(12): 1478-81.

[8]   Caterino JM. Evaluation and management of geriatric infections in the emergency department. *Emerg. Med. Clin. North Am.* 2008; 26(2): 319-43, viii.

[9]   Crossley KB, Peterson PK. Infections in the elderly. In: Mandel GL, Bennett JE, Dolin R, eds. Principles and practices of Infectious Diseases. 7 ed. Philadelphia: *Churchill livingstone*; 2010: 3865-75.

[10]   Iberti TJ, Bone RC, Balk R, Fein A, Perl TM, Wenzel RP. Are The criteria used to determine sepsis applicable for patients> 75 years of age? *Crit. Care Med.* 1993; 21: S130.

[11]   Chassagne P, Perol MB, Doucet J, et al. Is presentation of bacteremia in the elderly the same as in younger patients? *Am. J. Med.* 1996; 100(1): 65-70.

[12]   Meurer WJ, Smith BL, Losman ED, et al. Real-time identification of serious infection in geriatric patients using clinical information system surveillance. *J. Am. Geriatr. Soc.* 2009; 57(1): 40-5.

[13]   Khayr WF, CarMichael MJ, Dubanowich CS, Latif RH. Epidemiology of bacteremia in the geriatric population. *Am. J. Ther.* 2003; 10(2): 127-31.

[14]   Nee PA. Critical care in the emergency department: severe sepsis and septic shock. *Emerg. Med. J.* 2006; 23(9): 713-7.

[15]   Dellinger RP, Carlet JM, Masur H, et al. Surviving Sepsis Campaign guidelines for management of severe sepsis and septic shock. *Crit. Care Med.* 2004; 32(3): 858-73.

[16]   Garnacho-Montero J, Garcia-Garmendia JL, Barrero-Almodovar A, Jimenez-Jimenez FJ, Perez-Paredes C, Ortiz-Leyba C. Impact of adequate empirical antibiotic therapy on the outcome of patients admitted to the intensive care unit with sepsis. *Crit. Care Med.* 2003; 31(12): 2742-51.

[17]   Rivers E, Nguyen B, Havstad S, et al. Early goal-directed therapy in the treatment of severe sepsis and septic shock. *N. Engl. J. Med.* 2001; 345(19): 1368-77.

[18]   Nguyen HB, Rivers EP, Abrahamian FM, et al. Severe sepsis and septic shock: review of the literature and emergency department management guidelines. *Ann. Emerg. Med.* 2006; 48(1): 28-54.

[19]   Rajapakse S, Rajapakse A. Age bias in clinical trials in sepsis: how relevant are guidelines to older people? *J. Crit. Care* 2009; 24(4): 609-13.

[20] Sprung CL, Annane D, Keh D, et al. Hydrocortisone therapy for patients with septic shock. *N. Engl. J. Med.* 2008; 358(2): 111-24.

[21] Bernard GR, Vincent JL, Laterre PF, et al. Efficacy and safety of recombinant human activated protein C for severe sepsis. *N. Engl. J. Med.* 2001; 344(10): 699-709.

[22] Ely EW, Angus DC, Williams MD, Bates B, Qualy R, Bernard GR. Drotrecogin alfa (activated) treatment of older patients with severe sepsis. *Clin. Infect. Dis.* 2003; 37(2): 187-95.

[23] Alexander SL, Ernst FR. Use of drotrecogin alfa (activated) in older patients with severe sepsis. *Pharmacotherapy* 2006; 26(4): 533-8.

[24] van den Berghe G, Wouters P, Weekers F, et al. Intensive insulin therapy in the critically ill patients. *N. Engl. J. Med.* 2001; 345(19): 1359-67.

[25] Wiener RS, Wiener DC, Larson RJ. Benefits and risks of tight glucose control in critically ill adults: a meta-analysis. *JAMA* 2008; 300(8): 933-44.

[26] Adedipe A, Lowenstein R. Infectious emergencies in the elderly. *Emerg Med Clin North Am* 2006; 24(2): 433-48, viii.

[27] Erdem H, Kilic S, Coskun O, et al. Community-acquired acute bacterial meningitis in the elderly in Turkey. *Clin. Microbiol. Infect.* 2010; 16(8): 1223-9.

[28] Durand ML, Calderwood SB, Weber DJ, et al. Acute bacterial meningitis in adults. A review of 493 episodes. *N. Engl. J. Med.* 1993; 328(1): 21-8.

[29] Miller LG, Choi C. Meningitis in older patients: how to diagnose and treat a deadly infection. *Geriatrics* 1997; 52(8): 43-4, 7-50, 5.

[30] Behrman RE, Meyers BR, Mendelson MH, Sacks HS, Hirschman SZ. Central nervous system infections in the elderly. *Arch. Intern. Med.* 1989; 149(7): 1596-9.

[31] Gorse GJ, Thrupp LD, Nudleman KL, Wyle FA, Hawkins B, Cesario TC. Bacterial meningitis in the elderly. *Arch. Intern. Med.* 1984; 144(8): 1603-7.

[32] van de Beek D, de Gans J, Spanjaard L, Weisfelt M, Reitsma JB, Vermeulen M. Clinical features and prognostic factors in adults with bacterial meningitis. *N. Engl. J. Med.* 2004; 351(18): 1849-59.

[33] Choi C. Bacterial meningitis in aging adults. *Clin. Infect. Dis.* 2001; 33(8): 1380-5.

[34] Fitch MT, van de Beek D. Emergency diagnosis and treatment of adult meningitis. *Lancet Infect. Dis.* 2007; 7(3): 191-200.

[35] Talan DA, Guterman JJ, Overturf GD, Singer C, Hoffman JR, Lambert B. Analysis of emergency department management of suspected bacterial meningitis. *Ann. Emerg. Med.* 1989; 18(8): 856-62.

[36] Tunkel AR, Hartman BJ, Kaplan SL, et al. Practice guidelines for the management of bacterial meningitis. *Clin. Infect. Dis.* 2004; 39(9): 1267-84.

[37] van de Beek D, de Gans J, McIntyre P, Prasad K. Steroids in adults with acute bacterial meningitis: a systematic review. *Lancet Infect. Dis.* 2004; 4(3): 139-43.

[38] de Gans J, van de Beek D. Dexamethasone in adults with bacterial meningitis. *N. Engl. J. Med.* 2002; 347(20): 1549-56.

[39] van de Beek D, de Gans J, Tunkel AR, Wijdicks EF. Community-acquired bacterial meningitis in adults. *N. Engl. J. Med.* 2006; 354(1): 44-53.

[40] van de Beek D, de Gans J. Dexamethasone in adults with community-acquired bacterial meningitis. *Drugs* 2006; 66(4): 415-27.

[41] Monto AS, Ansaldi F, Aspinall R, et al. Influenza control in the 21st century: Optimizing protection of older adults. *Vaccine* 2009; 27(37): 5043-53.

[42] Clark NM, Lynch JP, 3rd. Influenza: epidemiology, clinical features, therapy, and prevention. *Semin. Respir. Crit. Care Med.* 2011; 32(4): 373-92.

[43] Van Kerkhove MD, Vandemaele KA, Shinde V, et al. Risk factors for severe outcomes following 2009 influenza A (H1N1) infection: a global pooled analysis. *PLoS Med* 2011; 8(7): e1001053.

[44] Nickel KB, Marsden-Haug N, Lofy KH, et al. Age as an independent risk factor for intensive care unit admission or death due to 2009 pandemic influenza A (H1N1) virus infection. *Public Health Rep.* 2011; 126(3): 349-53.

[45] Thompson WW, Shay DK, Weintraub E, et al. Influenza-associated hospitalizations in the United States. *JAMA* 2004; 292(11): 1333-40.

[46] Thompson WW, Shay DK, Weintraub E, et al. Mortality associated with influenza and respiratory syncytial virus in the United States. *JAMA* 2003; 289(2): 179-86.

[47] McElhaney JE. The unmet need in the elderly: designing new influenza vaccines for older adults. *Vaccine* 2005; 23 Suppl 1: S10-25.

[48] Fiore AE, Uyeki TM, Broder K, et al. Prevention and control of influenza with vaccines: recommendations of the Advisory Committee

on Immunization Practices (ACIP), 2010. *MMWR Recomm Rep* 2010; 59(RR-8): 1-62.

[49] Uyeki TM, Prasad R, Vukotich C, et al. Low sensitivity of rapid diagnostic test for influenza. *Clin. Infect. Dis.* 2009; 48(9): e89-92.

[50] Hurt AC, Alexander R, Hibbert J, Deed N, Barr IG. Performance of six influenza rapid tests in detecting human influenza in clinical specimens. *J. Clin. Virol.* 2007; 39(2): 132-5.

[51] Fiore AE, Fry A, Shay D, Gubareva L, Bresee JS, Uyeki TM. Antiviral agents for the treatment and chemoprophylaxis of influenza --- recommendations of the Advisory Committee on Immunization Practices (ACIP). *MMWR Recomm Rep.* 2011; 60(1): 1-24.

[52] Goodwin K, Viboud C, Simonsen L. Antibody response to influenza vaccination in the elderly: a quantitative review. *Vaccine* 2006; 24(8): 1159-69.

[53] Djuretic T, Ryan MJ, Fleming DM, Wall PG. Infectious intestinal disease in elderly people. *Commun. Dis. Rep. CDR Rev.* 1996; 6(8): R107-12.

[54] Garibaldi RA. Residential care and the elderly: the burden of infection. *J. Hosp. Infect.* 1999; 43 Suppl: S9-18.

[55] Andrews CN, Raboud J, Kassen BO, Enns R. Clostridium difficile-associated diarrhea: predictors of severity in patients presenting to the emergency department. *Can. J. Gastroenterol.* 2003; 17(6): 369-73.

[56] Kirk MD, Veitch MG, Hall GV. Gastroenteritis and food-borne disease in elderly people living in long-term care. *Clin. Infect. Dis.* 2010; 50(3): 397-404.

[57] Wilcox MH. Cleaning up Clostridium difficile infection. *Lancet* 1996; 348(9030): 767-8.

[58] Ricciardi R, Rothenberger DA, Madoff RD, Baxter NN. Increasing prevalence and severity of Clostridium difficile colitis in hospitalized patients in the United States. *Arch. Surg.* 2007; 142(7): 624-31; discussion 31.

[59] Vugia DJ, Samuel M, Farley MM, et al. Invasive Salmonella infections in the United States, FoodNet, 1996-1999: incidence, serotype distribution, and outcome. *Clin. Infect. Dis.* 2004; 38 Suppl 3: S149-56.

[60] Naumova EN, Egorov AI, Morris RD, Griffiths JK. The elderly and waterborne Cryptosporidium infection: gastroenteritis hospitalizations before and during the 1993 Milwaukee outbreak. *Emerg. Infect. Dis.* 2003; 9(4): 418-25.

[61] Danna PL, Urban C, Bellin E, Rahal JJ. Role of candida in pathogenesis of antibiotic-associated diarrhoea in elderly inpatients. *Lancet* 1991; 337(8740): 511-4.

[62] Trinh C, Prabhakar K. Diarrheal diseases in the elderly. *Clin. Geriatr. Med.* 2007; 23(4): 833-5.

Index

#

21st century, 120

A

Abraham, 15
access, ix, 65, 66, 67
accounting, 36, 58
acetaminophen, 6
acetic acid, 7
achlorhydria, 116
acid, 5, 7, 33, 34, 46, 50, 87
ADA, 46, 47
adalimumab, 103
adhesion, 86, 98
adolescents, 62, 79
adults, viii, xi, 19, 28, 31, 32, 34, 35, 37, 55, 61, 62, 99, 109, 112, 113, 114, 116, 119, 120
adverse effects, 6, 10, 75, 85, 89, 92, 117
adverse event, 54
aetiology, 80
African Americans, 91
age, xi, 19, 21, 23, 37, 38, 42, 58, 60, 109, 110, 112, 113, 114, 115, 116, 117, 118
agglutination, 87
agglutination test, 87

AIDS, 33, 62
airways, 111
albumin, 8
allergic reaction, 36
allergy, 116
alveolar macrophage, 34, 106
aminoglycosides, 50
anatomy, 113
anemia, 6
aneurysm, 83
angiography, 44
antibiotic, ix, 4, 6, 18, 20, 21, 28, 66, 68, 70, 71, 72, 74, 77, 79, 80, 112, 113, 114, 116, 117, 118, 122
antibiotic prophylaxis, x, 66, 72, 79
antibody, 3, 48, 84, 85, 98, 108
anticardiolipin, 98
antigen, 4, 61, 110
antihistamines, 7, 9
anti-inflammatory drugs, 11, 82
antimicrobial therapy, 4, 53, 57
antiphospholipid antibodies, 97
antituberculosis, ix, 32, 61, 62
antiviral drugs, 115
apathy, 37
aphasia, 37, 58
arabinoside, 14
arachnoiditis, 37, 39, 42

arrhythmias, 112
artery(ies), 42, 44, 56, 63
arthritis, 88, 89
arthrodesis, 20
aseptic, vii, viii, 1, 2, 3, 4, 5, 6, 7, 8, 9, 10,
    11, 12, 13, 14, 15, 68, 75, 85, 86, 87,
    90, 91, 93, 100, 104
aseptic meningitis, vii, viii, 1, 2, 3, 4, 5, 6,
    7, 8, 9, 10, 11, 12, 13, 14, 15, 68, 75,
    85, 86, 87, 90, 91, 93, 100, 104
aspartate, 99
assessment, 102
asthenia, 37
asymptomatic, 35, 44
ataxia, 37
atherosclerosis, 86
atrophy, 41
autoantibodies, x, 81, 86, 90
autoimmune disease(s), vii, x, xi, 6, 81,
    82, 85, 86, 90, 93
autoimmunity, x, 82
autopsy, 21, 44, 45, 88, 103, 105
Azathioprine, 5, 14

bleeding, 75
blindness, 102
blood, ix, xi, 6, 18, 21, 22, 24, 25, 34, 65,
    67, 68, 70, 73, 84, 86, 88, 95, 98, 109,
    110, 111, 112, 113
blood circulation, 34
blood cultures, xi, 18, 21, 22, 24, 25, 68,
    70, 109
blood flow, 112
blood stream, 34, 110, 111
blood vessels, 88
blood-brain barrier, ix, 65, 67, 70, 98
bloodstream, ix, 65, 67
bone, 68, 71, 74
brain, viii, ix, 6, 18, 20, 22, 27, 32, 34, 35,
    36, 38, 39, 43, 44, 46, 47, 50, 51, 52,
    56, 63, 65, 66, 67, 72, 73, 74, 77, 78,
    82, 86, 88, 92, 98
brain abscess, 20, 63
brain stem, 36, 43
brainstem, 43, 83
breakdown, 6, 68
bronchospasm, 116

B

C

bacillus, 32, 34, 35, 36, 37, 42, 46, 87
back pain, viii, 18
bacteremia, vii, xi, 34, 109, 110, 111,
    117, 118
bacteria, ix, 21, 24, 25, 27, 34, 66, 67, 70,
    73, 87, 111
bacterial infection, 19, 23, 69, 74
bacterium, viii, 18, 34
basal ganglia, 38, 44
base, 71
basilar artery, 44
BD, 83, 84, 85, 99
beneficial effect, 88
benefits, 33, 54, 71
bias, 28, 47, 75, 118
biochemistry, 18, 25
biopsy, 48, 88

calcification, 56
calcitonin, 23
candida, 122
capsule, 38, 44, 55
carbamazepine, 14
cardiovascular system, 115
carotid arteries, 44
catheter, 111
Caucasians, 91, 94
cauda equina, 43
causal relationship, vii, 2
cefazolin, 6
central nervous system (CNS), ix, x, 2, 4,
    10, 22, 32, 33, 35, 37, 46, 47, 59, 60,
    63, 65, 67, 69, 77, 82, 83, 84, 85, 86,
    90, 91, 93, 96, 98, 102, 104, 105, 106,
    107, 108, 114

cephalosporin, 4, 8, 21, 71, 73
cerebellum, 43, 55
cerebral arteries, 44
cerebral blood flow, 21
cerebral edema, 36, 38
cerebral tuberculomas, 56
cerebrospinal fluid, 2, 10, 18, 28, 61, 62, 67, 73, 79, 82, 95, 104, 106, 107, 114, 115
cerebrovascular disease, 97
cerebrum, 55
challenges, viii, 2
chemicals, 9
chemokines, 29
chemotherapeutic agent, 7
chemotherapy, 56, 57, 61, 101
childhood, 33
children, viii, 14, 19, 28, 31, 32, 34, 35, 36, 37, 38, 44, 45, 52, 55, 60, 78, 79, 115
chorea, 36, 37, 38
choroid, 5
Christians, 96
classes, 3
classification, 29, 74
clinical diagnosis, viii, 2
clinical examination, vii, 2
clinical presentation, 35, 70, 105, 110
clinical syndrome, 98
clinical trials, x, 66, 75, 77, 118
closure, 111
cognitive dysfunction, 86, 87, 90, 98, 99
cognitive function, 98
cognitive impairment, 37
colitis, 117, 121
Colombia, 65
colonization, ix, 65, 67, 68, 117
coma, viii, 18, 22, 27, 37, 38, 114
commercial, 6, 47
common signs, viii, 4, 18
common symptoms, 83
communication, 68

community, viii, 18, 21, 24, 74, 111, 114, 115, 120
compilation, vii
complement, 3, 4, 86
complications, viii, 18, 20, 25, 27, 48, 60, 66, 72, 73, 83, 84, 86, 88, 90, 92, 94, 96, 103, 105, 112, 115
compression, 42
computed tomography, 20, 90
conditioning, 43
conflict, 78
conjunctivitis, 4
connective tissue, 11, 90, 93, 99, 107
consciousness, viii, 18, 22, 24, 27, 36, 57, 69, 89, 111
consensus, ix, 22, 32, 50, 52, 93
contamination, 68
control group, 76, 77
controlled studies, 57
controversial, x, 33, 54, 66, 72, 74, 86, 89
correlation(s), 24, 38, 60, 63, 94, 103
corticosteroid therapy, 14
corticosteroids, 7, 14, 45, 54, 57, 82, 84, 85, 87, 89, 91, 99, 113
cough, 111, 115
cranial nerve, 36, 37, 43, 88, 90
craniotomy, 74, 75, 78, 80
cranium, 42
creatinine, 115
crises, 37
CRP, 22, 26, 27
CSA, 96
CT, ix, 14, 20, 21, 23, 32, 38, 39, 48, 51, 56, 73, 76, 90, 92, 102
CT scan, 20, 21, 48, 51, 56
cultivation, 18
culture, 3, 21, 22, 33, 34, 46, 47, 48, 67, 68, 74, 75, 76, 111
culture medium, 34, 46
cure, 90
cyclophosphamide, x, 82, 84, 85, 88, 89, 93, 96, 99, 101, 102, 104
cyclosporine, 89, 91, 93, 96, 106

cytokines, x, 81, 86
cytosine, 14
cytotoxic agents, 100

D

database, 19
death rate, 33, 116
deaths, 33
defects, 71, 86
deficiency, 86
deficit, viii, 18, 22, 36, 37, 38, 39, 48, 55, 58
delirium, 36, 37, 89, 110
dementia, 37
demyelinization, 38
dental abscess, 20
deposition, 5
depression, 86, 87, 99
destruction, 90
detection, ix, 11, 47, 48, 61, 65, 113, 115
developed countries, 32
developing countries, viii, 31, 32, 73
diabetes, 99, 106
diabetes insipidus, 106
diabetic patients, 113
diagnostic criteria, 74, 77, 91, 107
diagnostic markers, 110
dialysis, 113
diarrhea, vii, 116, 117, 121
diarrheas, 117
differential diagnosis, x, 9, 54, 82, 87, 93, 94
diffusion, 89, 103
diplopia, 4
disability, 54, 57, 73
discharges, 37
discomfort, 35, 36
disease activity, 84, 92, 99
diseases, x, xi, 3, 4, 78, 81, 82, 83, 93, 99, 110, 116, 122
disorder, 11, 38, 39, 111
disseminated tuberculosis, 38

distribution, 38, 43, 92, 121
DNA, 47, 48, 61
DOI, 62
donors, 6
dopamine, 112
dosage, 4, 92
drainage, 79
drug history, 8
drug interaction, 62
drug reactions, 13, 54
drug resistance, 52
drug withdrawal, 87
drugs, vii, viii, ix, 1, 2, 3, 7, 8, 9, 10, 11, 31, 32, 50, 52, 53, 54, 61, 84, 89, 93, 112, 115, 117
dura mater, 88
dystonia, 36, 37, 38, 59

E

ecchymosis, 73
Ecuador, 31
edema, 38, 42, 43, 52, 55, 56, 57, 58
egg, 116
elderly population, 110, 111, 112, 113, 115, 116, 117
elders, 115
ELISA, 98
emergency, xi, 109, 117, 118, 120, 121
empyema, 21
encephalitis, 12, 36, 43
encephalomyelitis, 39
encephalopathy, 38, 60
endothelial cells, 6, 86
England, 116
enlargement, 63
entrapment, 42, 51
enzyme, 48, 91, 106
eosinophils, 6, 7
epidemiology, 94, 117, 120
epinephrine, 112
equilibrium, 71

Escherichia coli, viii, xi, 18, 25, 70, 109, 112
etanercept, 85, 91
etiology, 38, 91, 112, 117
Europe, 83
evidence, x, 2, 3, 11, 23, 34, 66, 69, 73, 76, 77, 84, 86, 92, 93, 117
evolution, viii, 18
exclusion, x, 2, 8, 82
exposure, 6, 8, 9, 71
extensor, 39
extraction, 21
exudate, 36, 38, 43, 44, 58
eye movement, 37

F

fever, vii, viii, x, 2, 18, 24, 27, 35, 36, 37, 57, 69, 81, 87, 90, 110, 115
fibrosis, 88
Filipino, 99
firearms, 73
fixation, 4
flora, 67, 73, 112
fluid, 6, 7, 8, 10, 14, 23, 45, 79, 89, 95, 112, 117
fluoroquinolones, 50
food, 121
force, 100
Ford, 97
formation, 3, 35, 43, 93
fractures, 68, 71, 73, 74, 76, 77, 79
fragments, 68, 74
France, 11
frontal lobe, 37
fungi, 117
fusion, 98

G

gadolinium, 7, 88, 106
gait, 37

gastroenteritis, 117, 121
genome, 34
Germany, 1
glucose, 3, 21, 45, 57, 68, 87, 113, 119
glutamate, 86
glycine, 6
glycol, 6
gram negative organisms, 71
granulomas, 34, 40, 41, 42, 43, 47, 48, 49, 51, 52, 55, 92, 93
gray matter, 56
growth, 35, 45, 46, 56
guidelines, 61, 80, 113, 114, 118, 120

H

head injury, 74, 78, 79
head trauma, 78
headache, vii, viii, x, 2, 8, 18, 36, 37, 57, 58, 81, 83, 87, 90, 95
health, 28, 71
health care, 71
hearing loss, 73
heart transplantation, 102
hematomas, 38
hemiparesis, 37, 89
hemorrhage, 38, 68
hepatitis, 8, 14, 54
hepatotoxicity, 62
highly active antiretroviral therapy, 53
history, 8, 22, 35, 53, 113
HIV, viii, ix, 15, 31, 32, 33, 34, 35, 37, 42, 46, 47, 53, 56, 58, 61, 63
HIV infection, ix, 32, 42
HLA, 83, 94
homes, 116
hospitalization, 70, 71, 73, 115
host, 35, 37, 42, 66
human, 59, 61, 62, 119, 121
human immunodeficiency virus, 59, 61, 62
humoral immunity, 110

hydrocephalus, 32, 36, 38, 42, 43, 44, 48, 52, 54, 58, 77, 88
hydrocortisone, 7, 14
hypersensitivity, vii, 1, 3, 4, 6, 10, 12, 35, 42, 56
hypertension, 36, 55, 56, 58, 83, 87
hypoglycemia, 113
hypogonadism, 106
hypotension, 4, 23, 88, 112
hypoxemia, 110

I

ibuprofen, 2, 4, 9, 10, 11, 87, 100
identification, 46, 73, 110, 118
idiopathic, 13, 83, 89, 93
idiopathic thrombocytopenic purpura, 13
image(s), ix, 32, 38, 47, 52, 56, 57, 89, 92, 103
immune function, 57
immune response, 5, 35
immune system, 86, 110, 116, 117
immunity, 34, 87, 110
immunization, 71
immunocompromised, 20
immunoglobulin(s), 6, 9, 10, 12, 13, 91
immunomodulatory, 84
immunosuppression, 15
immunosuppressive, x, 82, 84, 87, 104
immunosuppressive agent, x, 82
impairments, 58
improvements, 34, 112
incidence, vii, 1, 2, 13, 33, 35, 54, 66, 67, 70, 73, 74, 75, 77, 97, 117, 121
incubation period, 3
India, 48
indium, 7
individuals, 3, 23, 32, 110, 115, 116, 117
inducer, 91
induction, 88
induction period, 88
infants, 36
infarction, 44, 60, 63

infection, vii, ix, xi, 2, 4, 8, 20, 22, 23, 32, 33, 34, 35, 37, 42, 61, 65, 66, 67, 70, 71, 73, 74, 76, 78, 82, 87, 91, 92, 93, 99, 109, 111, 112, 115, 117, 118, 119, 120, 121
infectious agents, 66, 87
inflammation, vii, 1, 2, 3, 10, 12, 29, 32, 42, 44, 73
inflammatory cells, 55, 88
inflammatory disease, 88
infliximab, 84, 85, 90, 91, 93, 96, 102, 107
influenza, vii, 114, 115, 116, 120, 121
influenza a, 114, 115, 120
influenza vaccine, 116, 120
ingestion, vii, 2, 3
inhibitor, 15
injections, 14
injury(ies), ix, 65, 66, 67, 70, 71, 73, 78
insulin, 113, 119
integrity, 86
intensive care unit, 22, 28, 118, 120
intervention, 67
intracranial aneurysm, 83
intracranial pressure, 20, 21, 37, 42
intrathecal, vii, 1, 3, 5, 6, 7, 9, 11, 12, 13, 14, 86
intravenous fluids, 112
intravenous immunoglobulins, vii, 1
intravenously, 75
inversion, 89
inversion recovery, 89
Iran, 76, 78
Iraq, 94, 95
irritability, 36, 37
ischemia, 36, 38, 58
isolation, 21, 67, 76, 117
isoniazid, 12, 50, 52, 53
issues, 33

J

Japan, 81, 83, 105

joints, 88

K

Korea, 100

L

laboratory tests, 87
lactic acid, 10
lactic acid level, 10
latency, 35
lead, 20, 54, 69, 73, 87
leakage, 67, 72, 79
leaks, x, 66, 68, 72, 74
lesions, 34, 38, 42, 43, 56, 57, 66, 84, 89, 92
lethargy, 4
leukocytes, 45, 90
Listeria monocytogenes, 87, 114
liver, 54, 112
liver function tests, 54
longitudinal study, 98
loss of consciousness, 55
lumbar puncture, 8, 18, 20, 21
lumen, 34
lupus, 10, 15, 86, 97, 98, 101
lymph, 42, 91
lymph node, 42, 91
lymphocytes, 43, 45, 87, 99

M

macrophages, 35
magnetic resonance, 7, 82, 96, 98, 106
magnetic resonance imaging, 7, 82, 96, 98, 106
malaise, vii, 2, 84
malignancy, 82, 92
malnutrition, 110
maltose, 6

management, 3, 7, 9, 59, 71, 94, 97, 100, 105, 111, 112, 114, 117, 118, 120
manic, 14
manic-depressive illness, 14
mass, 20, 42, 43, 57, 59, 112
mastoid, 68, 74
matrix, 45, 55, 62, 98
matrix metalloproteinase, 55, 62, 98
MB, 62, 79, 97, 103, 104, 118
mean arterial pressure, 112
measles, 8, 14
measurement, 23, 69
mechanical ventilation, viii, 18, 22, 25, 27, 113
media, 7, 13
median, 23, 24
medical, ix, 8, 19, 28, 32, 56, 68, 115
medication, 3, 9, 53, 82
Mediterranean, 83
medulla, 45
mellitus, 113
meninges, vii, 1, 3, 32, 34, 42, 43, 51, 92
meningitidis, viii, 18, 24, 25, 27, 114
mental state, 37, 69, 89
mental status changes, vii, 2
mesencephalon, 46
meta-analysis, 47, 61, 73, 76, 80, 119
methodology, 77
methylprednisolone, 13, 84
mice, 100
microbiological agents, ix, 66, 70
microemulsion, 96
microorganism(s), x, 66, 67, 68, 70, 77, 112, 116, 117
microscopy, 48
migration, 7
miliary tuberculosis, 34, 42
military, 33
molecules, 86
monoclonal antibodies, x, 82, 91
monoclonal antibody, 5, 93, 96
morbidity, ix, 18, 50, 52, 65, 66, 83, 94, 113, 116

mortality, ix, xi, 18, 21, 22, 23, 24, 27, 28, 50, 52, 57, 58, 65, 66, 72, 75, 77, 83, 94, 98, 109, 112, 113, 114, 115, 116
mortality rate, xi, 24, 109, 114, 115, 116
Moses, 105
movement disorders, 36, 38
MR, 13, 52, 94, 98, 101, 102, 103, 117
MRI, ix, 32, 38, 39, 40, 41, 44, 46, 47, 48, 50, 51, 56, 57, 62, 82, 88, 89, 91, 92, 94, 101, 103
multiple sclerosis, 13, 90, 92, 93, 107
multivariate analysis, 63
mumps, 8, 14
myalgia, 115
myasthenia gravis, 13
mycobacteria, 34
Mycobacterium tuberculosis, ix, 32, 61, 87
myocardial infarction, 113
myoclonus, 36, 37
myopathy, 106

N

natural killer cell, 110
nausea, vii, x, 2, 81, 87, 90
necrosis, 43, 45, 55, 88, 96
nephritis, 101
nerve, 6, 37, 88
nervous system, 2, 14, 59, 67, 82, 88, 90, 91, 94, 95, 96, 99, 104, 105, 106, 107, 119
neuroimaging, 56, 68, 83
neurological disability, 54
neurological impairment, viii, 31, 58
neurologists, ix, 32
neuromyelitis optica, 93, 107
neurons, 90
neuropathy, 92
neurosarcoidosis, 85, 91, 105, 106, 107
neurosurgery, 69, 74
neurotoxicity, 15, 84, 96

neutrophils, 99, 110
NMDA receptors, 99
nodes, 43
nodules, 88
non-steroidal anti-inflammatory drugs, 11
North America, 83, 91
NSAIDs, vii, 1, 3, 4, 5, 6, 9, 82, 86, 87
nuchal rigidity, vii, 2, 36, 114
nuclei, 33, 34
nucleic acid, 48, 61
nucleus, 38
null, 77
nursing, 116
nursing home, 116
nystagmus, 37

O

obstruction, 42, 43
occlusion, 45
oedema, 4
optic chiasm, 42, 43, 44
optic nerve, 42, 88
optic neuritis, 101, 102
oral cavity, 74
organ(s), x, 22, 23, 28, 33, 81, 86, 88, 90, 110, 112
organism, 34, 53
osteoarthritis, 114
otorrhea, 73
oxygen, 112

P

pain, viii, 18, 36, 39
pancreatitis, 2, 4, 9
papilledema, 36, 37
paralysis, 73
parasitic diseases, 56
parenchyma, 32, 34, 68, 72, 92
parkinsonism, 36, 37
parotitis, 4

pathogenesis, viii, x, 3, 5, 31, 33, 34, 38, 59, 60, 81, 86, 93, 122

pathogens, 67, 74, 114

pathology, 60, 62, 88, 90

PCR, 47, 48, 61, 87

PCT, 19, 22, 23, 26, 27

penicillin, 12, 73, 75, 80, 114

perfusion, 112

pericarditis, 20

peripheral nervous system, 90

personality, 37

pertussis, 8

petecchiae, viii, 18

PGE, 60

pH, 3

pharmacokinetics, 112

phenotypes, 107

Philadelphia, 118

photophobia, viii, 18

physicians, 3, 110, 116

physiology, 28

pia mater, 89

placebo, 54, 76, 80

plasma cells, 43

plasmapheresis, 105

plexus, 5

pneumococcus, viii, 18, 25, 27, 71

pneumonia, viii, 18, 22, 24, 67, 72

polymerase, 47, 61

polymerase chain reaction, 47, 61

population, 32, 91, 96, 110, 112, 113, 114, 116, 117, 118

posttraumatic meningitis, vii, x, 66, 67, 69, 70, 71, 73, 74, 76, 77, 80

prednisone, 104

preparation, 7

prevention, 15, 34, 74, 76, 80, 120

prevention of infection, 74

primary care physician, vii, viii, 1, 2, 3

principles, 54

probability, 19, 22, 35, 114

procalcitonin, viii, 18, 19, 27, 28, 69

prognosis, viii, ix, 21, 31, 32, 39, 44, 48, 50, 54, 57, 58, 66, 72, 84, 86, 93, 115, 116

pro-inflammatory, 86

proliferation, 55

prophylactic, vii, x, 66, 67, 70, 71, 72, 74, 75, 76, 77, 80

prophylaxis, x, 66, 72, 79, 80

protection, 33, 120

proteins, ix, 21, 32, 37, 38, 45, 53, 56, 98

pseudomembranous colitis, 116

Pseudomonas aeruginosa, 70

psychosis, 86, 87

pus, 57

pyogenic, 36, 57

R

randomized controlled clinical trials, 76

RE, 101, 119

reactants, 27

reactions, 3, 8, 11, 13

recommendations, 97, 100, 120, 121

recovery, 48, 58, 87

recurrence, 8

reflexes, 39

reintroduction, 54

rejection, 10, 15

relapses, 86, 92

reliability, 92

relief, 91

remission, 89

renal failure, 6, 22

repair, 71, 79

researchers, 75

Residential, 121

resistance, viii, ix, 31, 32, 48, 71, 112, 115

resolution, vii, 2, 3, 8, 53, 56, 68, 73

respiratory syncytial virus, 120

response, 37, 39, 42, 53, 56, 68, 69, 91, 104, 107, 108, 111, 121

restoration, 62, 112

RH, 14, 118
rhabdomyolysis, viii, 18, 22, 25
rheumatoid arthritis, 2, 3, 85, 101, 102, 103
rheumatoid factor, 88
rhinorrhea, 73, 115
ribosome, 86
risk(s), ix, x, 6, 8, 13, 28, 32, 45, 66, 67, 68, 71, 72, 73, 74, 75, 76, 80, 84, 87, 99, 110, 112, 113, 115, 116, 119, 120
risk factors, 13, 67, 73, 80, 110, 113, 116
rituximab, 85, 88, 91, 101
roots, 36, 43, 57
rotavirus, 116
rubella, 8, 14

S

safety, 71, 119
salivary gland(s), 90
Salmonella, 116, 117, 121
sarcoidosis, 2, 56, 85, 91, 92, 93, 105, 106, 107
saturation, 112
sediments, 48
seizure, viii, 18
sensitivity, 6, 28, 33, 39, 47, 48, 68, 83, 111, 115, 121
sepsis, vii, xi, 23, 28, 109, 110, 111, 112, 113, 117, 118, 119
septic shock, viii, 18, 22, 23, 25, 27, 112, 113, 118, 119
serum, 3, 4, 28, 69, 90, 91, 98
services, ix, 65, 66
shock, 27, 112
showing, 75
side effects, x, 13, 66, 75, 77, 115
signs, vii, ix, 2, 7, 24, 32, 33, 36, 37, 38, 51, 53, 57, 66, 73, 82, 87, 89, 90, 92, 110
sinuses, ix, 65, 67, 68, 73, 74
sinusitis, 20, 67
skin, 42, 68, 84, 88, 91, 111
skull fracture, 71, 73, 75, 76, 79, 80
sodium, 14
solubility, 3, 7
somnolence, 36
South Africa, 62
Spain, 17
species, 70
sphincter, 37
spinal anesthesia, 5, 14
spinal cord, 32, 36, 40, 43, 57
spinal stenosis, 20
spleen, 86
spore, 34
Spring, 12
SS, 90
SSA, 104
standard deviation, 23, 27
starch, 7, 14
state(s), 36, 69, 111
stenosis, 44
sterile, 23
steroids, x, 39, 55, 61, 82, 87, 99, 104, 113, 115
stimulation, 96
stratification, 28
streptococci, 111
Streptococcus, viii, ix, 18, 25, 66, 70, 111, 114
stress, 87, 110
striatum, 50
stroke, 20, 43, 45, 86
stupor, 38, 113
subacute, 36, 37, 39, 56, 57, 90, 92
subarachnoid space, ix, 34, 35, 36, 37, 42, 55, 57, 65, 67, 73, 89
sucrose, 6
surgical intervention, 71
surgical removal, 57
surveillance, 79, 118
survival, 18, 35, 54, 112, 113
survival rate, 113
survivors, 72
susceptibility, 42, 71, 80

Switzerland, 61

symptoms, vii, viii, ix, x, 1, 2, 3, 4, 6, 7, 8, 9, 18, 22, 24, 25, 27, 32, 33, 35, 36, 37, 53, 66, 69, 81, 82, 83, 87, 88, 90, 91, 92, 93, 96, 110, 114, 115, 116

syndrome, 2, 6, 10, 12, 15, 23, 53, 62, 83, 85, 90, 94, 95, 96, 103, 104, 105, 116

synthesis, 6, 11, 90

syringomyelia, 37

systemic lupus erythematosus, x, 2, 3, 8, 10, 11, 12, 14, 81, 85, 97, 98, 99, 100, 101

T

T cell, 3, 8, 86

tachycardia, 110

tachypnea, 110

target, 33, 56

TBI, ix, 65, 66, 67, 68, 69, 70, 71, 73, 74, 75, 76, 77

TBM, viii, ix, 31, 32, 33, 34, 35, 37, 38, 39, 42, 43, 44, 45, 46, 47, 48, 50, 52, 53, 54, 56, 57, 58

tension, 84, 87

territory, 43

testing, 47

tetanus, 8

textbook, 13

thalamus, 44

therapy, 3, 9, 10, 12, 13, 14, 15, 42, 45, 50, 53, 54, 56, 57, 62, 63, 71, 77, 84, 85, 87, 92, 96, 97, 118, 119, 120

third-generation cephalosporin, 73

thrombosis, 44, 58, 83, 88

tissue, 3, 4, 35, 43, 44, 55, 74, 77, 89, 112

tissue perfusion, 112

TNF, 84, 89, 96, 97, 107

TNF-alpha, 107

TNF-α, 89

toxicity, 7, 14, 42, 53, 62

toxoplasmosis, 56

trajectory, 74

transfusion, 112

transient ischemic attack, 88

transmission, 33, 34

transplant, 9

transplantation, 85, 88, 100, 101

transport, 34

trauma, ix, 34, 65, 66, 67, 68, 69, 71, 74, 75, 78, 79

traumatic brain injury, 73

tremor, 36, 37, 38

trial, 54, 80, 101

tuberculosis, viii, 31, 32, 33, 34, 35, 38, 39, 45, 50, 53, 55, 57, 59, 60, 61, 62, 63, 87

tuberculous meningitis, ix, 32, 33, 42, 43, 59, 60, 61, 62, 63

tumo(u)r(s), 2, 84, 106

tumor necrosis factor, 84, 106

Turkey, 109, 119

U

UK, 14, 83

uncal herniation, 20

United States, 33, 117, 120, 121

urinary tract, viii, xi, 18, 22, 109, 110

urinary tract infection, viii, 18, 22, 110

usual dose, 115

uveitis, 83, 84, 95, 96

V

vaccinations, 8

vaccine, 8, 14, 71, 116

vancomycin, 21, 71, 114

variables, 21, 22

vascular occlusion, 36

vasculature, 6

vasculitides, 107

vasculitis, 5, 36, 38, 42, 49, 51, 55, 56, 90, 93, 96, 102, 105

vessels, 38, 43, 44

Vietnam, 54
viral infection, 8
viral meningitis, 28
virus infection, 120
viruses, 2
vomiting, vii, viii, 2, 18, 36, 37, 87, 110

W

weakness, 39, 75, 110
wealth, 34
weight loss, 37
white blood cell count, 84
white matter, 39, 92
withdrawal, 9

World Health Organization(WHO), 33,
    48, 53, 61
World War I, 75
worldwide, 34

X

xerophthalmia, 90
xerostomia, 90

Y

yield, 46
young adults, 115
young people, 38